Introduction

Japan was not slow in recognizing the potential of naval aviation. In June 1912 a Naval Aeronautical Research Committee was established by the Imperial Japanese Navy, and shortly afterwards a number of naval officers were sent to France and the United States with orders to purchase several seaplanes for training purposes. Four aircraft – two Glenn Curtiss and two Maurice Farman seaplanes – were acquired, and in November that year a small naval air base was set up at Yokosuka, on the coast, enabling a cadre of naval officers to receive flight and technical training there.

Following the purchase of more aircraft in 1913, the Imperial Japanese Navy commissioned its first seaplane tender, the *Wakamiya Maru*, converted from a transport vessel, and in September and October 1914 four Maurice Farmans operating from this ship carried out bombing and reconnaissance operations against the German fortress of Tsingtao, on the Chinese coast. By the end of the First World War, two Naval Air Corps, the Yokosuka and Sasebo Kokutais, had been activated, the first in April 1916 and the second in March 1918.

The two years after the end of the war saw little progress in the development of Japanese naval aviation, although in June 1920, Lieutenant Kuwabara Torao, who had been attached to the Royal Navy in 1917 and who had reported on the development of Britain's early aircraft carriers, made the first successful take-off by a Japanese naval aviator from a platform mounted on the *Wakamiya Maru*. He was flying an imported Sopwith Pup.

The early 1920s saw the establishment of Kawasaki, Mitsubishi and Nakajima as the 'big three' of Japan's embryo aircraft industry, whose early development work relied heavily on aid from Britain, France, Germany and the USA. Japanese factories concentrated mainly on the licence production of foreign designs and on the overhaul of types purchased directly from abroad, such as the Gloster Sparrowhawk (a development of the Nieuport Nighthawk).

A key participant in Mitsubishi's development work was Herbert Smith, who had been chief engineer with the Sopwith

A Japanese postcard showing a Mitsubishi B1M attack aircraft with Mount Fuji in the background. The type entered service in 1924 and served into the 1930s.

Two Maurice Farman MF11 seaplanes were purchased by the Imperial Japanese Navy in 1912 and used to train a cadre of naval aviators. (via Armand Agababian)

Top: The Nakajima A1N2 was based on the British Gloster Gambet. It entered service with the IJN in 1930.

Above: Japan's first aircraft carrier, the *Hosho*, was completed in 1922. (via J.R. Cavanagh)

Aviation Company and as such had led the team that had designed the famous Pup, Triplane, Camel and Snipe. In February 1921, the Sopwith company having been dissolved, the Mitsubishi Internal Combustion Engine Manufacturing Company in Nagoya invited Smith, along with several other former Sopwith engineers, to assist Mitsubishi in setting up an aircraft manufacturing division. Smith and his team were instrumental in designing the Mitsubishi B1M, the first Japanese aircraft designed for the torpedo attack role, and the Mitsubishi 1MF, a carrier-borne biplane fighter. In February 1923 it made the first successful take-off by a Japanese-built aircraft from Japan's first aircraft carrier, the *Hosho*. Production of the 1MF ended in 1929, after 128 had been built.

The 1MF was replaced by a Nakajima fighter, the A1N, which was a licence-built version of the British Gloster Gambet, itself a carrier-borne derivative of the earlier Gloster Gamecock. Fifty aircraft were built as the A1N1 and were followed by the A1N2, an improved version that entered service in 1930. About 100 aircraft were produced by 1932, and on 22 February that year A1Ns from the carriers *Hosho* and *Kaga* scored the Imperial Japanese Navy's first air-to-air combat victory when they shot down a Chinese Air Force Boeing P-12 during the so-called Shanghai Incident. The P-12 was flown by an American volunteer pilot, Robert Short.

By the mid-1930s, naval aviation in Japan was gathering momentum. In 1935 Nakajima produced the A4N1, which entered service as the Type 95 Carrier Fighter; 221 were built and the type saw action in the Sino-Japanese conflict of 1937, operating in both the air superiority and ground attack roles. Meanwhile, in 1934, a Naval Aircraft Establishment had been set up to oversee the development of new types under a programme called 9-Shi. This referred to the ninth year of Showa, the reign of His Imperial Highness Hirohito, and it was to see significant progress on the road that would eventually lead to the Zero fighter.

Japanese aircraft designations

The Imperial Japanese Navy Air Service (IJNAS) used several different aircraft designation systems simultaneously. Between 1931 and 1945, aircraft were allocated Shi (experimental) numbers denoting the specification to which they were designed. They also had a long form of type and model number system used between 1920 and 1943, a short designation system, and a system of popular names introduced to replace type numbers from 1943 to 1945 (for example, Reppu – Hurricane).

After 1929, aircraft types were given a type number based on the last two digits of the Japanese imperial year (which is counted from the mythical founding of Japan in 660BC by the Emperor Jimmu). Added to this was a brief description of the aircraft's function. The Mitsubishi Zero was so called because it entered service in 1940, which was the Japanese year 2600, and so it was designated Type 0 Carrier Fighter. A short designation scheme was also adopted. This used a letter or two letters to designate the type of aircraft, a number to indicate the number in series of that type of aircraft, and finally a letter to designate the manufacturer. Therefore, the Zero's designation was A6M, which denoted that it was the sixth type of carrier fighter designed under this designation system, and that it was built by Mitsubishi. Variants developed for other roles were indicated by a dash, followed by an additional letter; (for example, A6M2-N).

Left: The Nakajima A4N1 entered service in 1935 as the Type 95 Carrier Fighter, and saw action in the Sino-Japanese war of 1937. (via J.R. Cavanagh)

Below: The crowded deck of the aircraft carrier *Kaga*, pictured in 1937. The aircraft on deck are Nakajima A2Ns, Aichi D1As and Mitsubishi B2Ms. (Kure Maritime Museum)

The aircraft carrier *Akagi* seen leaving Osaka with her air group ranged on deck. (Kure Maritime Museum)

The Japanese Carrier Fleet 1922–1942

In the 1930s, the true state of the Japanese Navy and its associated maritime air strength was unknown to the western powers. Great Britain had played a considerable part in assisting Japan to build a modern fleet in the early years of the century, but that association had come to an end with the rise of Japanese militarism, and in the 1930s Japan's naval and military activities had become effectively hidden from the outside world. Naval air power was at the core of Japan's strategy. She had completed her first aircraft carrier, the *Hosho*, in 1922, and this vessel had soon been followed by larger and more powerful vessels like the 26,000-ton *Kaga*. Launched on 17 November 1920, *Kaga* was originally laid down as an improved *Nagato*-type dreadnought, with increased armour protection and an enlarged main battery. She was cancelled in February 1922 to comply with the terms of the Washington Naval Treaty, but instead of scrapping her, the Japanese Naval Staff decided to complete her as an aircraft carrier to replace another carrier, the *Amagi*, which had been destroyed in an earthquake while under construction. *Kaga* was completed in March 1928 and joined the Combined Fleet in 1930. In 1934/5 she was reconstructed, a full-length flight deck and island being added. Recommissioned in June 1935, she was assigned to the First Carrier Division with the *Akagi*, her air group seeing action during the Sino-Japanese war. Japanese fast-attack carrier construction up to 1939 culminated in the 25,675-ton *Shokaku* and *Zuikaku*. After mid-1942, the Japanese were unable to make good the losses sustained at Midway and elsewhere, and although 21 new carriers were planned, only the *Taiho* and three Unryu-class vessels were completed, all of them to be lost in the battles of 1944.

The Monoplane Fighters

First flown in prototype form in January 1935, the Mitsubishi A5M was Japan's first carrier-borne monoplane fighter. Its appearance was of great importance, for it marked the end of Japanese dependence on foreign designs. The prototype had an inverted gull wing, but subsequent aircraft featured a wing with a straight centre section and dihedral on the outboard panels. The initial production model was designated A5M1 Type 96 and had an enclosed cockpit, the first to be used by a Japanese fighter. It was not popular with the A5M's pilots and subsequent variants reverted to an open cockpit. These were the A5M2a, with a more powerful engine, and the A5M2b, with a three-blade propeller. The A5M was armed with two fixed forward-firing .303in (7.7mm) machine guns in the upper forward fuselage, and it could carry an external bomb load of 132lb (60kg).

In August 1937, during the conflict between Japan and China, six A5M1s were assigned to the aircraft carrier *Kaga*, and on 22 August two of them were flown to an airstrip near the front in Manchuria, making their first sortie that same day. On this occasion no contact was made with the enemy, but on 4 September two Type 96s, led by Lt Chikamasa Igarashi, shot down

The aircraft carrier *Kaga* seen travelling at speed in 1930. (Kure Maritime Museum)

The Mitsubishi A5M Type 96 was the first of the Imperial Japanese Navy's monoplane fighters. The A5M2b, seen here, was fitted with a three-blade propeller. (Kure Maritime Museum)

The British Gloster Gladiator, seen here in RAF insignia, was a match for the A5M, but suffered from poor pilot training in China. (BAe Systems)

The Imperial Japanese Army's equivalent of the A5M was the Nakajima KI.27, which remained the army's standard fighter until mid-1942.

three Curtiss Hawks. On the 7th, three aircraft led by Lt Igarashi claimed to have destroyed several Chinese aircraft, Igarashi claiming three of them. From then on the *Kaga*'s Type 96 fighters were widely used in China. In one air battle against 21 Gloster Gladiators, Curtiss Hawks and other types, which took place on 30 August 1938, the *Kaga* fighter squadron claimed to have destroyed eleven enemy aircraft for the loss of two A5Ms.

The Imperial Japanese Army Air Force's equivalent of the Navy's Mitsubishi A5M, the Nakajima Ki.27, first flew in October 1936 and was to remain the JAAF's standard fighter up to the middle of 1942, playing a prominent part in Japan's campaigns in China and South-East Asia. Production totalled 3,495 aircraft, the main variants being the Ki-27a with an uprated engine and a metal-faired canopy, and the Ki.27b with a clear-view canopy and underwing bomb racks. The type was used mainly as a fighter trainer after 1942. Armament was similar to that of the A5M, although the aircraft could carry a heavier bomb load of 220lb (100kg).

The Monoplane Fighters in Combat

It was during the Sino-Japanese conflicts of the 1930s that the Japanese fighter pilots of both army and navy developed the expertise and tactics that would make them such formidable opponents over the Pacific and Burma.

The A5M and Ki-27 had been in service for only a short time when their merits were put to the test in combat. In July 1937, the Japanese launched a full-scale invasion of China, quickly capturing urban centres along the Chinese coast and pushing rapidly along the Yangtse river as Chinese forces retreated westward. During the initial period of operations, the Imperial Japanese Army left the brunt of the air fighting to the Navy, limiting its own activity to air support of ground operations along the Manchurian border while new units were formed.

The IJN was well placed to conduct offensive air operations. Its first-line fighter squadrons were now equipped with Mitsubishi A5Ms, its carrier-based attack squadrons with the Yokosuka B4Y bomber and its land-based bomber squadrons with the twin-engine Mitsubishi G3M (later to be codenamed Nell by the Allies). In August 1937, following the conclusion of a non-aggression pact between China and the Soviet Union, the Soviet Central Committee agreed to supply quantities of aircraft and equipment for use by the Chinese Central Government. The first contingent of Soviet Air Force personnel arrived in China during the last week of September, and by the beginning of 1938 there were over 350 Russians serving in the country, a total that included 80 fighter pilots. Although their task was ostensibly to give flying instruction to Chinese aircrew converting to Russian types, their real purpose was to fly four squadrons of Polikarpov I-152 (I-15bis) fighters in combat against the Japanese. Two squadrons of SB-2 bombers also arrived at the same time, and they, too, were manned by Russian crews.

Operational flying began almost immediately from bases in the vicinity of Nanking, the Chinese capital, which by that time was coming under heavy Japanese air attack. In January 1938, two I-15 squadrons, supplied to the Chinese some time earlier, were pulled out of the front line and replaced by two fresh units operating I-16s, which proved able to meet the Japanese A5M fighters on more equal terms. Meanwhile, Chinese units had begun to receive the first of 400 I-152s and I-16s promised by the Soviet government, and Russian-trained Chinese pilots were flowing through the flying and technical schools at Nanchang, Langchow and Hami at a fast rate – so fast, in fact, that many of them were far from ready to take on the highly skilled Japanese aircrews in combat. Nevertheless, so desperate was the situation that they were immediately sent into action in the defence of Nanking, replacing the Soviet squadrons. That was in mid-November. Four weeks later, their pilots completely outclassed by the Japanese, the Chinese squadrons had been decimated and had to be withdrawn to Nanchang, leaving the capital vulnerable.

The Russian squadrons, meanwhile, had regrouped near Langchow, where a new batch of crews had arrived to start their four-month tour of duty in China. There was little air activity on either side during the spring of 1938, but in April and May Japanese reconnaissance aircraft began to appear in increasing numbers over Langchow and other places in north-west China, where the combat aircraft being supplied by Russia were assembled. The first heavy air raids on these centres were carried out late in May and continued for five weeks, heavy losses being inflicted on the attacking Mitsubishi G3M bombers – which had a very long range but poor

The Polikarpov I-16, seen here being serviced by Soviet naval personnel, was one of the fighter types supplied to China by the Soviet government. (TASS)

defensive armament – by the defending fighters.

Two months later, in July 1938, a major clash occurred between Soviet and Japanese forces after Russian troops occupied a strategic ridge near Lake Khasan on the Soviet–Manchurian border. The Japanese claimed that the ridge was theirs and attacked the Russian positions on 29 July, capturing the objective. During the bitter fighting that followed, four squadrons of TB-3 heavy bombers, escorted by I-152s, repeatedly pounded the Japanese positions. While Soviet infantry and tanks fought their way back towards the ridge, I-16s and I-152s strafed ahead of them. Japanese attacks the next day were also broken up by Soviet aircraft, about 180 of which were committed to the battle.

On 11 May 1939 the Japanese, anxious to avenge their defeat at Lake Khasan, launched a major offensive into Mongolian territory at Khalkhin Gol. There was sporadic aerial combat during the opening weeks of the battle, but the battles of August were to be marked by the biggest air confrontation the world had seen since the First World War. During July, the Japanese had assembled a total of 475 aircraft of all types in Manchuria, including considerable numbers of Mitsubishi A5M and Nakajima Ki-27 fighters. There were also some bomber squadrons equipped with Mitsubishi Ki-21s, fast twin-engine bombers which had entered service with the Japanese Army Air Force the previous year and which were later to be known by the Allied codename of Sally. The Russians, meanwhile, had been pouring troops and equipment into Mongolia, and the First Army Group – commanded by General Georgi K. Zhukov, who was to achieve fame during the Second World War – was rapidly deployed along the disputed frontier.

At his disposal, at the end of July, Zhukov had 580 aircraft operated by both Soviet and Mongolian squadrons. The total included 150 I-16s and 200 I-15s/152s. The I-15s had already clashed with Japanese fighters, and had proved inferior to the Nakajima Ki-27; during subsequent operations they were relegated mainly to the ground attack role, leaving the I-16s and I-152s to engage the enemy. During August, four of the Red Air Force squadrons exchanged their I-15s for the more modern I-153, with a retractable undercarriage, and the appearance of these aircraft over the Khalkhin Gol took the Japanese fighter pilots by surprise, especially as the Russians evolved tactics which were deliberately designed to mislead the enemy. They would approach the combat area with their undercarriages lowered, giving the impression that they were slower I-15s or I-152s and inviting the Japanese to attack. The Ki-27s suffered some losses before their pilots realized what was happening.

Aerial combat over the Khalkhin Gol sometimes involved as many as 200 aircraft. The Japanese usually emerged the worse from these encounters; although the Ki-27 could hold its own against the I-153 and was superior to the I-152, it was outclassed in speed and firepower by its most frequent opponent, the I-16, which the Japanese nicknamed Abu (Gadfly). Figures of the air losses sustained by both sides are vague and conflicting. The Russians were reported to have lost 145 aircraft of all types between 11 May and 15 September 1939, when a ceasefire was arranged. The Japanese are said to have lost 600, 200 of them during the last ten days of the fighting, but these figures seem over-optimistic.

Nakajima Ki-27 fighters at Nomonhan (Khalkhin Gol) in 1939. (via John McKenna)

What is certain is that very heavy aircraft losses were suffered by both sides, and particularly by the Japanese. Although there had not been a great deal of difference in terms of skill between Russian and Japanese pilots, the Russian tactics had been better. So had the armament: four machine guns mounted in both the I-16 and I-153, compared to the twin MGs carried by the Ki-27 and A5M. The Japanese Army Air Force's poor showing in the fighting against the Russians, and the clear superiority of Russian equipment and tactics, led the Japanese High Command to consider the USSR as the principal potential enemy, and equipment planning was influenced by the requirement to be prepared for a renewed conflict on the Manchuria–Siberia border. New combat aircraft were developed to carry out tactical missions in cold weather, so that they were poorly suited to long over-water missions among the Pacific islands in the coming war of 1941–45.

The Imperial Japanese Navy, on the other hand, bore the brunt of long-range operations against Chinese targets, the aircraft industry developing bombers for missions at extreme range and also a fighter capable of escorting them to the target and back. The result was one of the finest aircraft of all time, the Mitsubishi A6M Zero.

Mitsubishi Zero: Design & Development

On 19 May 1937, preliminary specifications for a new Navy Experimental Carrier Fighter to replace the A5M were submitted to Mitsubishi and Nakajima under the 12-Shi development programme. An analysis of the A5M's combat performance in China revealed a number of shortcomings, the most important of which was that the combat radius of the A5M was insufficient to escort Japanese bombers on deep-penetration missions to targets in mainland China. The specification for the new 12-Shi carrier fighter was therefore revised and made more stringent, not only because of the experience in China, but also because reports were reaching Japan of the new generation of fighter monoplanes under development in the West, such as the Messerschmitt Bf 109, the Hawker Hurricane and Supermarine Spitfire, and the Curtiss Hawk 75A.

The new specification, issued in October 1937, called for a fighter capable of intercepting and destroying all types of bomber, and of performing the escort mission with a performance better than that of other contemporary fighter aircraft. Maximum speed was to be better than 311mph (500km/h) at 13,123 feet (4,000m) in level flight, with the ability to climb to 9,843 feet (3,000m) within 3 minutes 30 seconds of leaving the ground (3min 45sec from start of the take-off roll). Endurance at 9,843 feet (3,000m) at normal power setting, fully loaded with auxiliary fuel tank, was to be between one and two hours, or up to eight hours at maximum range power setting. (In fact operational trials later indicated that an endurance of ten hours was attainable at 115 knots at 12,000 feet (3660m) and a power setting of 1,700 to 1,850rpm, reducing fuel consumption to 18 gallons per hour.) Take-off run in still-air conditions was not to exceed 574 feet (175m), reducing to 230 feet (70m) with a 30mph (43km/h) head wind. The new fighter was to be armed with two 20mm cannon and two .303in (7.7mm) machine guns, and there was to be provision for the carriage of two 66lb (30kg) or 132lb (60kg) bombs.

Following a technical conference held at the Naval Aircraft Establishment in Yokosuka on 17 January 1938, Nakajima

Cutaway drawing showing details of the Zero's relatively simple yet innovative structure. (Via John McKenna)

A Mitsubishi A6M Zero on its landing approach. (NASM)

decided that it was impossible to meet the demands of the revised specification, and dropped out of the competition. Mitsubishi elected to proceed, and appointed a talented young aero engineer called Jiro Horikoshi to assemble and lead a design team. Horikoshi had joined Mitsubishi in 1926, aged 23, and in June 1929 he had been sent on a tour of aircraft factories in England, France, Germany and the Netherlands, remaining in Europe for six months before leaving for the United States, where he spent several months attached to the Glenn L. Curtiss factory in Garden City, familiarizing himself with the technical aspects of the Curtiss P-6 pursuit aircraft, an example of which was about to be delivered to Japan. Subsequently, he led the team responsible for the design of the highly successful Mitsubishi A5M (Type 96), and he was now tasked with organizing a team to develop the new 12-Shi fighter. Key personnel included Yoshitoshi Sone and Teruo Tojo, responsible for the theoretical calculations; Yoshio Toshikawa, for the structural work; Denichiro Inoue and Shotaro Tanaka for the engine installation; Yoshimi Hatakenaka for the armament and ancillary equipment, which included the oxygen system, engine fire extinguisher, lighting equipment and instrumentation; and Sadahiko Kao and Takeyoshi Moro for the landing gear and associated equipment.

Saving weight was the key factor in the new design, and to this end Horikoshi's team incorporated a number of revolutionary features. The wing was built in one piece, eliminating the need for the heavy centre-section fitting normally required for joining two separate wing sections. The one-piece wing combined with the engine, cockpit and forward fuselage to form a single rigid unit, the centre section of the fuselage being riveted to the upper wing skin, which formed the cockpit floor. The rear fuselage, with the tail, was joined to the forward section by a series of 80 bolts, fixed to two ring formers just aft of the cockpit. The aircraft could therefore easily be split into two components, facilitating storage and also providing unobstructed access to the cockpit area from the rear.

Another weight-saving device was the use of a lightweight alloy called Extra-Super Duralumin (ESD) in the manufacture of the wing main spar. Manufactured by Sumitomo Metal Industry, it had a tensile strength 30 to 40 per cent higher than any previous alloy. Although very light in weight (3,704lb/1,680kg empty, compared with the Spitfire Mk I's 4,341lb/1,969kg), which gave an impression that the airframe was structurally weak, the 12-Shi fighter design was actually very strong.

Horikoshi and his design team had given much thought to the 12-Shi fighter's aerodynamics. The aircraft was designed to produce the minimum drag, while remaining inherently stable and easy to fly. Because of the requirement to reduce the wing loading to a figure less than 21.5lb/ft^2 (105kg/m^2) to achieve optimum take-off, climb and manoeuvrability, the aircraft was given a wing area of 241.54ft^2 (22.44m^2) and a new aerofoil section was selected. This was the Mitsubishi 118, which offered minimum centre of pressure movement, an important factor in reducing drag.

Powerplant

As no suitable liquid-cooled in-line engine was available in Japan, three 14-cylinder radials were evaluated: the 875hp Mitsubishi Zuisei (Holy Star) 13, the 950hp Nakajima Sakae (Prosperity) 12, and the 1,070hp Kinsei (Golden Star) 46. The Nakajima engine was initially rejected as it was designed by a competitor; Horikoshi favoured the Kinsei, but because of the Navy's insistence on a power loading not exceeding 5.5lb/hp, this heavier and more powerful engine was abandoned and it was the Zuisei 13 that was selected to power the prototype. A two-blade variable-pitch propeller was fitted to the prototype, but this was replaced during the flight test phase by a constant-speed three-blade Sumitomo-Hamilton propeller.

Armament

The primary armament selected for the 12-Shi fighter was the licence-produced 20mm Oerlikon cannon, a Swiss-designed weapon which was already used by a number

The Zero's clean, uncomplicated lines are well illustrated in this excellent shot of a captured A6M2. (NASM)

The two-blade propeller originally fitted to the A6M was soon replaced by the three-blade Hamilton one, seen here. (NASM)

of European fighter aircraft, including France's Morane M.S.406 and Dewoitine D.510. Japan had acquired two of the latter aircraft for evaluation in 1935, and results showed that although the Oerlikon had a relatively low muzzle velocity, it was light and sufficiently compact to permit installation in an aerodynamically clean wing. A further advantage was that it fired explosive shells. A new company, the Dai-Nihon Heiki (Japan Munitions Company) was established to mass-produce the weapon and its ammunition under licence as the Type 99 Standard Naval Aircraft Cannon. The early version used drum magazines, while the later Type 99 Model 2 Mk 4 was belt-fed. The weapon was constantly improved throughout the war, some 35,000 cannon being produced by half a dozen factories. In the 12-Shi fighter design, one Type 99 was installed in each wing just outboard of the landing gear, each gun having 60 rounds. The secondary armament comprised two Type 97 .303in (7.7mm) machine guns, installed in the upper forward fuselage above the engine and firing through the propeller arc.

The prototype 12-Shi Experimental Carrier-Borne fighter was rolled out at Mitsubishi's Nagoya factory on 16 March 1939, and three days later, after completion of weight and balance calculations and preliminary engine runs, it was dismantled and shipped on two ox-drawn wagons to the Army flying field at Kagamigahara, about 25 miles (40km) to the north. The aircraft was re-assembled and Mitsubishi test pilot Katsuzo Shima carried out taxi trials before making a short first flight – in reality, a brief hop along the runway to test the control responses – at 0730 on 1 April 1939. The flight test programme proper started the next day, and apart from some necessary minor adjustments to the landing gear no snags were encountered apart from a vibration problem, which disappeared when the two-blade propeller was replaced by the three-blade constant-speed one.

On 14 September 1939, having accumulated 119 hours of flight testing by Mitsubishi's company pilots and a further 43 with Navy pilots at the controls, the aircraft was officially accepted by the Imperial Japanese Navy and given the designation A6M1 Type 0 Carrier-Borne Fighter. The alpha numeric designation was seldom used, the fighter being commonly known as the Rei Shiki Sento Ki (Type Zero Fighter) often abbreviated to Rei-Sen or Reisen.

When it first appeared, many foreign observers at once claimed that the Zero was a copy, or at least an imitation, of western designs. Jiro Horikoshi, writing a decade later, was quick to challenge these assertions.

"In summing up the defence of the design originality of the Zero, I will give credit where credit is due. As virtually all competent aeroplane designers will agree, the business of creating any new aeroplane is a process of adapting the original art and science to the problem at hand. For example, I will state that the undercarriage retraction design on the Zero was inspired by the Vought 143, and that the system for fastening the engine cowl and the method of mounting the engine came from other foreign planes. But nothing else, as far as the airframe is concerned. It is no exaggeration to say that we did not look upon the general design or basic configuration of foreign planes with great respect.

"In the case of accessories, many of these were built under licence from abroad; wheels were manufactured by the Okomato Engineering Company under licence from Bendix and Palmer, instruments were built by the Tokyo Instrument Company under licence, or later in the war by direct copy from Sperry, Pioneer and Kollsman. Sumitomo built hydromatic propellers under a licence from Hamilton Standard, as well as the German VDM propeller. The Nihon Musical Instrument Co. built the Junkers and Schwarz propellers, while the Kogusi Aircraft Company built the French Ratier airscrew. We built 20mm cannon licensed by Oerlikon of Switzerland and copies of the .50in (13mm) Browning." (Jiro Horikoshi: 'I Designed the Zero', *Royal Air Force Flying Review*, Vol. 12 No. 4, December 1956.)

On 25 October 1939, the Navy took delivery of the second Zero prototype, which, among other tasks, undertook armament trials. The third prototype, which began flight trials on 28 December 1939, was powered by the Nakajima NK1C Sakae 12 engine, common sense having at last triumphed over inter-company rivalry. This was to become the first production model of the Zero. Designated A6M2, it was to provide the backbone of the Imperial Japanese Navy's fighter force during the first half of the Pacific war.

On 11 March 1940, the second prototype was carrying out a series of dives to investigate an engine overspeeding problem at Oppama airfield, Yokosuka, when the aircraft disintegrated, killing its pilot, Lt Okuyama. Okuyama was actually thrown clear and his parachute deployed, but he slipped from his harness and fell to his death in the sea. The cause of the accident was never fully established, although the chief suspect was elevator flutter after failure of the mass balance; this would have created severe vibration, imposing intolerable stresses on the airframe. Despite this setback, full production was initiated in November 1940, the first A6M2s being assigned to the Yokosuka Air Group Fighter Squadron after completing their production flight test programme at the Naval Air Technology Arsenal. The Air Group's task was to bring the new fighter into full operational squadron service, flying the aircraft to its limits and developing combat tactics to suit its superior performance.

The Zero Variants

Such, then, was the beginning of the Zero's development story, but there were several more steps to be taken before the road ended five years later. The original production batch of 64 aircraft, designated A6M2 Model 11, was followed by the Model 21 with folding wingtips, and this was the major production model at the time of the attack on Pearl Harbor in December 1941. In 1942 Japanese Navy fighter units began to receive the A6M3 Model 32, with a supercharged 1300 Sakae 21 engine. This model had its folding wingtips removed to improve performance, but this

The port wingtip of a restored Mitsubishi A6M3 Zero Carrier Fighter Model 22 tail number X-133 operated by the Commemorative Air Force, seen at the Pacific Coast Dream Machines' 2009 vehicle shown at the Half Moon Bay Airport in Half Moon Bay, California.

This A6M3 Model 32 (Zeke 32) was rebuilt from the salvaged parts of five Zeros abandoned at Buna, Papua New Guinea, and tested at Wright Field in the USA. (USAF)

The A6M5 (Zeke 52) version of the Zero was built in greater numbers than any other variant. Introduced in the late autumn of 1943, it was intended to counter the F6F Hellcat. (USAF)

The A6M7 Model 63 was a variant specially developed for kamikaze suicide attacks. This captured example was shipped to the US for testing. (NASM)

impaired the fighter's manoeuvrability and the full-span wing was restored in the A6M3 Model 22. By early 1943, it was becoming apparent that the A6M3 could no longer retain superiority over the latest Allied fighters, so the A6M5 Model 52 was developed, retaining the Sakae 21 engine but having a shorter-span wing (in essence, a Model 32 wing with rounded tips). Sub-types produced included the A6M5a Model 52A, with strengthened wings and increased ammunition, the A6M5b Model 52B with heavier armement and armour protection, and the A6M5c Model 52C, with more armour, two 20mm and three 13mm guns. The latter sub-type had a Sakae 31 engine with methanol injection, bullet-proof fuel tanks and underwing rocket rails. The A6M7 Model 63 was a fighter-bomber version, of which 465 were built. Other versions were the A6M8c Model 54C, with a 1500hp Mitsubishi Kinsei 62 engine and four wing guns; a twin-float seaplane, the A6M2-N; and the A6M2-K2 two-seat trainer. In all, 10,937 Zeros of all versions were built.

Specification: A6M2 Model 11

Crew: 1
Powerplant: one 950hp Nakajima NK1C Sakae 12 14-cylinder radial engine

The A6M7 Zero. Note the underwing attachments for anti-personnel bombs. (NASM)

Performance: Max speed 332mph (534km/h); service ceiling 32,810ft (10,000m); range 1,929 miles (3,104km)
Dimensions: Wing span 39ft 4in (12m); length 29ft 8in (9.06m); height 10ft (3.05 m)
Weight: 6,164lb (2,796kg) loaded
Armament: two 20mm (0.79in) cannon and two fixed 7.7mm (.303in) machine guns; external ordnance load of 265lb (120kg)

Japanese Naval Air Formations – Designations

At the outset of the Pacific war the Imperial Japanese Navy was under the supreme command of the Imperial General Headquarters (Daihonei). Subordinate to this was the Chief of Navy General Staff (Gunreibu Socho), and subordinate to him in turn was Navy Air Headquarters (Kaigun Koku Hombu), which was responsible for the acquisition of aircraft, engines and associated equipment, as well as overseeing all aircrew and technical personnel training. The Combined Fleet (Rengo Kantai) was responsible for all operational matters, and operational control of naval air units was the responsibility of the various fleets (Kantais), Air Fleets (Koku Kantais) and Area Fleets (Homen Katais). Carrier-based aircraft were assigned to specific Carrier Divisions (Koku Sentais), which were usually formed around a nucleus of two aircraft carriers.

The carrier-based air groups of the Imperial Japanese Navy were designated Kokutai, later renamed Koku Sentai, which could have a complement of up to 90 aircraft, depending on their mission or the size of the aircraft carrier on which they were based. Each naval air fleet was assigned one or more naval air flotillas, each with two or more Kokutai. Each of these comprised several squadrons (Hikotai) with nine, 12 or 16 aircraft; the Hikotai was the basic combat unit and was divided into sub-sections (Shotai), each

An A6M8 Zero, one of the two prototypes built, preparing for a test flight at Misawa air base with an American pilot at the controls. (NASM)

An American soldier poses in the cockpit of an abandoned A6M2-K2 trainer. (World War Photos)

with three or four aircraft. The IJN had over 90 air groups at the start of the war, most of which were land-based and designated Koku Kantai. The crews of the carrier-based Kokutai were the elite of the Imperial Japanese Navy, and the Zero pilots were the best of the best.

Combat over China

In July 1940, the Imperial Japanese Navy, keen to assess the performance of the Zero in action, ordered Mitsubishi to release six fighters from the test programme for deployment to the 12th Air Group Fighter Squadron in China. On 21 July, the aircraft arrived at Wuhan under the command of Lt Tamotsu Yokoyama to be joined at intervals by nine more during the following month.

On 19 August, Yokoyama led 12 Zeros on a mission to escort 54 Mitsubishi G3M2 bombers in an attack on Hankow. No enemy fighters appeared, and they were also absent on 12 September, when the Zeros escorted 27 bombers in a raid on Chungking. On this occasion the Japanese fighter pilots carried out strafing attacks on airfields in the vicinity as the bombers left the target area.

The first opportunity for air combat came the next day, when 13 Zeros led by Lt Saburo Shindo provided the escort for another raid on Chungking. No enemy fighters appeared during the actual attack, but as the bombers and their escort turned for home a reconnaissance aircraft reported that 30 fighters were assembling over Chungking, apparently preparing to pursue the retreating bombers. Shindo at once ordered his fighters to turn back towards the target, climbing hard as they did so. The Chinese fighters – Soviet-built Polikarpov I-15 biplanes and I-16 monoplanes – were taken by surprise, and the air battle that followed was one-sided in the extreme. The Japanese pilots claimed the destruction of no fewer than 27 enemy aircraft, either shot down or damaged to such an extent that they crashed while

Right: Lt Saburo Shindo pictured with his dog, with Zero fighters in the background. Shindo survived the Pacific war and died in 2000.

Far right: Saburo Sakai was acknowledged as the leading tactician of the Tainan Air Group.

Pilots of the 12th Air Group Fighter Squadron seen in 1940. Several of them became air aces; most later lost their lives.

attempting to land. This battle saw the emergence of the first Zero air ace: he was Warrant Officer Koshiro Yamashita, who shot down five Chinese fighters. Other successful pilots were Petty Officer 2nd Class Yoshio Oki with four victories; PO 1st Class Tora-Ichi Takatsuka, with three; six other pilots destroyed two each, while three – one of whom was Shindo – claimed one each.

The appearance of the Zero came as an unpleasant shock to the Chinese. Their pilots, poorly trained to start with, began to regard the fighter with almost superstitious awe. On 16 September, 1940, three days after the Chungking battle, six Zeros pounced on a large, unidentified Chinese aircraft and shot it down. This was the last mission flown in September, the Zeros being temporarily withdrawn for inspection and overhaul in preparatory missions that would take them even deeper into Chinese territory. In this respect the Japanese were a long way in advance of other air forces in that the Zeros were fitted with long-range fuel tanks, installed under the fuselage. This was at a time when the Battle of Britain was raging in English skies, and the Luftwaffe's Messerschmitt Bf 109s were constantly hampered by lack of endurance in their fighter sweeps and escort missions over England; but the prowess of the Zero and its successes in combat over China went unnoticed in the West.

Operational flying resumed on 4 October 1940, when eight Zeros accompanied 27 bombers on a long-range mission to Chengtsu, in the extreme west of Szechwan Province. No defending fighters were encountered, but after leaving the target area the Zero pilots made a surprise attack on Taipingssu airfield, shooting down five I-16s and a Tupolev SB-2 bomber that were in the vicinity before strafing the airfield itself. Air reconnaissance subsequently revealed that at least 19 enemy aircraft had been destroyed on the ground.

The Zero fighter quickly established total Japanese air superiority in the China theatre, enabling the bombers to fly deep-penetration missions with virtual impunity, their crews secure in the knowledge that the Zero escorts were more than capable of dealing with anything the enemy might put up against them. Between 19 August 1940 and the end of the year, the Zeros flew 153 individual sorties in the course of 22 missions, destroying 59 Chinese aircraft in air combat and 101 on the ground for no loss to themselves.

In October 1940 the Mitsubishi G3M2 bomber force was withdrawn from operations in China to be reorganized and re-equipped, as it had suffered considerable losses prior to the arrival of the Zeros. For the next six months, until the bombers returned to China in April 1941, the Zeros conducted the air war virtually single-handedly, and with their extra fuel tanks they had sufficient combat

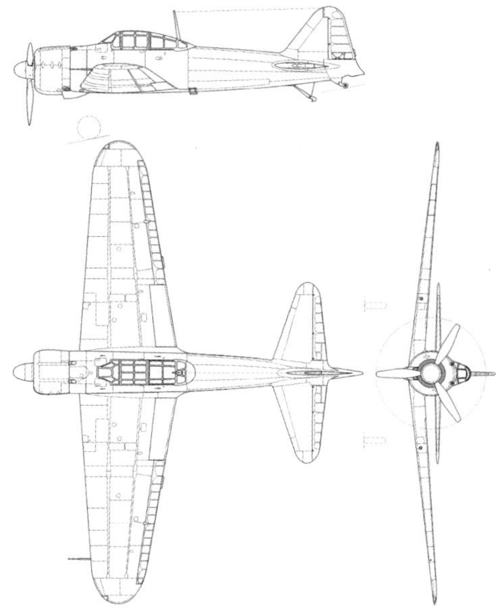

Three-view drawing of the Mitsubishi A6M Zero. (Kaboldy)

range to attack the Chinese on all fronts. Between 1 January and 1 September 1941 the Zeros flew 354 sorties, in the course of which they destroyed 44 enemy aircraft and damaged 62 more for the loss of two of their own number, both brought down by anti-aircraft fire. The 12th Air Group disbanded in September 1941, most of its pilots being transferred to the Tainan Air Group or the 3rd Air Group.

The 12th Air Group was not the only Zero unit to see combat in China. On 14 July 1940, the 14th Air Group Fighter Squadron, which had been supporting land operations in South China, moved to Hanoi in French Indo-China, the northern part of which had been occupied by Japanese forces following France's defeat by Germany in June 1940. Although mainly equipped with A5M Type 96 aircraft, the fighter element of this formation included nine Zeros, and on the day of their arrival at Hanoi seven were detailed to escort 27 ground attack aircraft of the 15th Air Group in an attack on Kunming, in south-western China. Over the target, the Japanese raiders were intercepted by a number of I-153, I-16 and Curtiss Hawk fighters; in an air battle lasting 15 minutes, 13 of them were shot down by the Zeros. Four enemy aircraft were also destroyed on the ground. And on 12 December, seven Zeros, with army reconnaissance aircraft leading the way, flew the 340 miles (547km) to Siangyun airfield and destroyed 22 Chinese aircraft in strafing attacks.

The 3rd Air Group also operated out of Hanoi in the summer of 1941. Apart from some reconnaissance aircraft, it was an all-fighter formation; this was unique among Japanese air groups, which were normally composed of a mixture of fighters, army co-operation and bomber aircraft. Many of its pilots had been transferred from the 12th Air Group, and were among the most experienced in the world. Even the newest pilots had around 1,000 hours' flying time to their credit.

Preparing for War

In mid-October the bulk of the 3rd Air Group moved to Takao, Formosa (Taiwan), with 45 A6M2 Zeros and 12 A5Ms. A further 13 of the Group's Zeros were detached to the 22nd Air Flotilla HQ in southern French Indo-China. The other Formosa-based air formation was the Tainan Air Group, which was to become best-known of all naval air groups and which was to boast the largest number of Japanese naval fighter aces. Formed at Tainan on 1 October 1941, the Group formed part of the 23rd Air Flotilla and, like the 3rd Group, had an establishment of 45 Zeros and 12 A5Ms. Fourteen of its Zeros were attached to the 22nd Air Flotilla in Indo-China in preparation for operations in Malaya and the Dutch East Indies. Taking into account the detached 3rd Air Group fighters, the 22nd Air Flotilla had 26 Zeros on strength, plus some A5Ms, and on 1 December 1941 these deployed forward to Saigon.

Meanwhile, the Formosa-based Zero units had embarked on a period of very intensive training as a prelude to the Japanese invasion of the Philippines. The original attack plan had envisaged the use of three small aircraft carriers – the *Ryujo*, *Zuiho* and *Taiho* – to bring the Zeros close to their objectives, but these vessels were limited in their operational capability and it was realized that a better option would be for the Zeros to fly directly to their main objective, Clark Field in the Philippines, and return non-stop.

Saburo Sakai, later to become one of the Imperial Japanese Navy's leading and best-known Zero aces, explained the problems.

"Clark Field was 450 miles away from our own air base, and Nichols Field, another major target near Manila, was 500 miles distant from Tainan. That meant, considering the factors of still-air range, fuel for fighting, and fuel for reserve, that we would be required to fly non-stop for some 1,000 to 1,200 miles! No fighters had ever flown on such combat missions before, and there were vehement arguments among the air staff as to whether even the Zero was capable of this performance ...

"We flew literally day and night to stretch the range of our planes ... the Zero was designed to remain in the air on a single flight for a maximum of six or seven hours. We stretched this from ten to 12 hours, and did so on mass formation flights. I personally established the record low consumption of less than 17 gallons per hour; on the average our pilots reduced their consumption from 35 gallons per hour to only eighteen. The Zero carried a normal fuel load of some 182 gallons.

"To conserve fuel, we cruised at only 115 knots at 12,000 feet altitude. Under normal full-power conditions, the Zero was capable of 275 knots and, when overboosted for short emergencies, could reach its maximum speed of about 300 knots. On our long-range flights we lowered propeller revolutions to somewhere between 1,700 and 1,850 revs per minute, and throttled the air control valve to its leanest mixture. This furnished us the absolute minimum of power and speed, and we hung on the fringe of losing engine power at any time and stalling.

"These new long-range cruising methods extended the Zero's range by a remarkable figure, however." (Saburo Sakai and Martin Caidin, *Samurai!*, William Kimber, London, 1959.)

'Battleship Row' at Pearl Harbor, seen from a Japanese aircraft during the attack. (*Times of Japan*)

The Pacific Campaign December 1941–January 1943

Final preparations for the Japanese attack on Pearl Harbor began on 22 November 1941. The fleet that assembled in Hitokappu Bay, in the remote Kurile Islands off the northern tip of Japan, was the most formidable assembled in Japanese waters since Tsushima, 36 years earlier. At its core were the aircraft carriers of the 1st Air Fleet, the *Akagi*, *Kaga*, *Hiryu*, *Soryu*, *Zuikaku* and *Shokaku*, under the command of Vice-Admiral Chuichi Nagumo. They were accompanied by the battleships *Hiei* and *Kirishima*, the heavy cruisers *Tone* and *Chikuma*, and the light cruiser and nine destroyers of the 1st Destroyer Squadron.

This was the battle fleet assigned by the Imperial Japanese Navy to what was known as the Hawaiian Operation, the planned surprise attack on Pearl Harbor, the principal base of the US Pacific Fleet. The fleet sailed on 26 November 1941, still prepared to turn back if diplomatic negotiations between Japan and the United States reached a satisfactory conclusion. The did not, and the coded signal Niitaka Yama Nobore (Climb Mount Niitaka), signifying that the Hawaiian Operation was to go ahead, was transmitted to Vice-Admiral Nagumo on 2 December. After replenishment at sea, the fleet proceeded to its flying-off position 370km (200NM) north of Oahu.

The first strike, launched at dawn on 7 December 1941, was made by 40 Nakajima B5N torpedo-bombers led by Lt Cdr Shigeharu Murata, 49 B5Ns carrying armour-piercing bombs and led by Commander Mitsuo Fuchida, and 51 Aichi D3A1 dive-bombers under Lt Cdr Kakwichi Takahashi. It was escorted by 43 Mitsubishi A6M Zero fighters of the first wave covering force under Lt Cdr Shigeru Itaya, officer commanding the Akagi Fighter Squadron.

Itaya's Zeros arrived over Pearl Harbor at 0750 Honolulu time; after shooting down a light aircraft and three trainers, they strafed Hickam and Ewa airfields, counting some 25 US aircraft burning on the ground. Itaya's wingman, PO 1st Class Takashi Hirano, was shot down by anti-aircraft fire. Six Zeros from the Shokaku and six from the Zuikaku strafed Kaneohe and Bellows fields, setting on fire 33 aircraft, many of them PBY amphibians. These two carriers did not take part in the second wave attack on Pearl Harbor, their fighters instead flying combat air patrols (CAPs) over the fleet in relays after refuelling and re-arming.

The first attack lasted 30 minutes, striking the anchorage and outlying air bases. The second strike, just over an hour later, was made by 54 *Kate* bombers, 78 D3A1 dive-bombers and 35 fighters; this lasted 65 minutes but was hampered by dense smoke from the burning anchorage and by heavy anti-aircraft fire, as well as

Surrounded by smoke and flame, the USS *California* is seen ablaze and sinking. (US Navy)

Airfield installations on fire at Wheeler Field, on the island of O'ahu. (NASM)

Tail section of a Zero shot down over Pearl Harbor, bearing the markings of the Akagi Fighter Squadron. (National Archives)

Clark Field in the Philippines seen in 1939, with American aircraft parked in neat rows. They were still parked in much the same fashion when the Japanese attacked in December 1941. (US Navy)

by small numbers of American fighters. Of the 94 warships in the harbour, 18 were sunk or suffered major damage; eight were battleships, the attackers' primary targets. During the second attack, nine Zeros from the Akagi led by the highly experienced Lt Saburo Shindo were detailed to escort 18 Aichi D3A1 dive-bombers. Arriving over the target about an hour after the first wave, and finding no opposition, the Zeros strafed Hickam Field, but only succeeded in destroying two aircraft on the ground.

The carrier *Kaga* despatched nine Zeros under Lt Nikaido, shooting down one aircraft and destroying about 20 aircraft on the ground, but four of the Zeros were shot down by US fighters. It was the worst loss sustained by any of the carrier fighter squadrons. Nine Zeros from the *Soryu* under Lt Fusato Iida also encountered strong opposition from American fighters during the second attack. Iida himself was shot down and killed in a strafing attack on Kaneohe, in which six PBYs were destroyed. Lt (Jg) Iyozo Fujita took over and was leading the flight away from the target area when they were intercepted. The Japanese pilots claimed two US fighters, but lost two of their own in the action. Eight Zeros from the *Hiryu* strafed Kaneohe and Bellows fields, their attack cut short when they became involved in an air battle with US fighters, two of which were shot down by PO 1st Class Tsugio Matsuyama. One Zero pilot, PO 1st Class Shigenori Nishikaichi, made a forced landing on the Hawaiian island of Niihau, and was killed by a Hawaiian. In all, the Zero squadrons lost 15 aircraft out of a total of 79 deployed during the Hawaiian operation; of the 384 US aircraft deployed on the islands, 188 were destroyed and a further 159 damaged.

Eighteen of the 94 US warships in Pearl Harbor were sunk or suffered major damage. Among the casualties was the battleship *Arizona*, pride of the Pacific Fleet and flagship of Rear Admiral Isaac C. Kidd, which was hit by a torpedo and eight bombs, exploded and sank with the loss of 1,404 lives. Also hit was the *California*. Severely damaged by bombs and with 98 of her crew dead, she sank three days later. A similar fate befell the *Nevada*, *Oklahoma*, *West Virginia*, *Maryland*, *Pennsylvania* and *Tennessee*, all of which were sunk, beached, or, at the very least, heavily damaged, leaving the Pacific Fleet crippled in the space of just a few hours.

The three aircraft carriers of the US Pacific Fleet were not present at the time of the attack. The USS *Saratoga* (CV-3) was just out of overhaul and was moored at San Diego; the USS *Lexington* (CV-2) was at sea about 680km (425mi) southeast of Midway Island, towards which she was heading to deliver a Marine scout bombing squadron; and the USS *Enterprise* (CV-6) was also at sea about 320km (200mi) west of Pearl Harbor, returning from Wake Island after delivering a Marine fighter squadron. It was these carriers that, in the early months of 1942, would form the nucleus of a rebuilt Pacific Fleet.

On Formosa, the attack on Pearl Harbor was received with wild jubilation by the Japanese fighter pilots, whose own mission had been delayed by unexpected fog. It was not until four hours after the Pearl Harbor raid that the fog cleared and the pilots were at last able to get airborne, 51 Zeros of the 3rd Air Group and 44 of the Tainan Air Group making the long over-water flight to the Philippines to attack Iba and Clark airfields on Luzon. Numerous American aircraft were destroyed on the ground, and about 19 in air combat. The 3rd Air Group lost two Zeros and the Tainan Air Group one, plus four that went missing on the return flight. On 10 December, both air groups were involved in a major air battle near Manila, claiming 44 Allied aircraft destroyed and 42 damaged. Attacks on airfields in the vicinity of Manila were repeated on the next three days, but there was hardly any opposition and it was clear that the Allied air forces in the Philippines had been virtually wiped out. Two USAAF Pursuit Groups, the 24th and 35th, were involved in these battles; they were armed with the Curtiss P-40, with the exception of the 34th Pursuit Squadron of the 35th Pursuit Group, which had Curtiss P-35As. The latter group had only received its new fighters on 7 December.

One of the P-40 units in the thick of the fighting was the 17th Pursuit Squadron, whose P-40s were based at Del Carmen alongside those of the 34th Pursuit Squadron. The 17th Squadron's commander was Lieutenant Boyd D. 'Buzz' Wagner, who on 11 December was leading a flight of P-40s on patrol north of Aparri when they were attacked by five Zeros. In two days of air fighting Wagner had learned how to cope with the agile Japanese fighters. Now, as two Zeros dived on him, he waited until they were in firing range and then suddenly throttled back. Taken by surprise, the Zero pilots shot past his aircraft. Wagner opened the throttle again, quickly caught up with them in the dive and shot both of them down. A week later, on 18 December, Wagner destroyed three more enemy aircraft to become the first US Army Air Forces ace of the Second World War. A few days later he was almost blinded when a cannon shell exploded in his cockpit, filling his eyes with Perspex splinters, but he brought his aircraft home and recovered to fly P-40s and P-39s in New Guinea, raising his score to eight. He was killed in a flying accident later in the war.

The P-40 squadrons fought on until February 1942; then, with the battle for the Philippines lost and their aircraft destroyed, the surviving P-40 pilots evacuated to Australia, where they joined the 49th Fighter Group at Darwin in time to provide the principal fighter defence of the vital northern Australian port. The 49th was equipped with a more modern version of the Warhawk, the P-40E. On 3 April 1942, one of these was delivered to Darwin by Lt John D. Landers, who was in action the next day, shooting down an enemy bomber. Almost immediately afterwards, he was pounced on by several Zeros, who shot his P-40 full of holes. He dived away and shook them off, then climbed over Melville Island in the hope of picking off any Japanese stragglers returning from the battle over Darwin. A while later, three Zeros passed directly under him and he dived on them, shooting one down. They were the first of six victories Landers was to score during his tour with the 49th. Later, flying with the Eighth Air Force in Europe, he added eight and a half more kills to his total.

Lieutenant James B. Morehead had already destroyed two Ki 21 bombers over Java before joining the 49th FG in March 1942. On 29 April 1942, ANZAC Day, the Japanese launched a major attack on Darwin and all three squadrons of the 49th Group were scrambled to intercept the incoming bombers and their fighter escort.

Lt Boyd 'Buzz' Wagner fought the Zeros over the Philippines to become the first US Army Air Forces air ace of the Second World War. (National Archives)

Morehead, leading the 8th Squadron, shot down three Mitsubishi G4M bombers, closing right in to make certain of his kills and braving very heavy return fire. His P-40 was hit by 39 machine-gun bullets and two cannon shells, which damaged it so badly that Morehead had to make a belly landing back at base. During his tour with the 49th, Morehead shot down two Zeros over Darwin, bringing his total to seven. He later added a Messerschmitt 109 to his tally while serving with the 1st Fighter Group in Italy. He retired from the USAF after the war with the rank of colonel.

On 14 December 1941, a flight of the Tainan Air Group moved into Legaspi airfield in the Philippines, captured by Japanese forces, and continued to provide cover for the ground forces until the end of the year, when the whole Group of 41 Zeros was ordered to deploy to the remote island of Horo. The move was completed by 7 January 1942, by which time the Group had begun attacks on the island of Trakan, Borneo, in preparation for an amphibious assault. By 11 January, the island had been captured, and the Group began its move forward on the 16th. Meanwhile, the 3rd Air Group had moved up to Davao, Mindanao, on 23 December 1941, from where it was also able to conduct operations over Tarakan. On the 28th, seven Zero fighters engaged nine Dutch Brewster Buffalo fighters, destroying almost all of them. On 12 January, the 3rd Air Group began moving up to Kendari, in the Celebes, which had been overrun, and flew attack missions to Kupang, on Timor.

On 24 January 1942, the Tainan Air Group provided cover from its Tarakan base for Japanese Army landings at Balikpapan, Borneo. In the course of these operations the Zero pilots encountered a variety of Allied bombers, including the B-17 Flying

The Brewster Buffalo, seen here in the markings of the Royal Netherlands East Indies Army Air Force, suffered heavily in combat against the Zero. (Netherlands National Military Museum, Soesterberg)

Non-commissioned fighter pilots of the 2nd Squadron, Tainan Wing, Lae, June 1942. Saburo Sakai is in the middle row, second left. (NASM)

Fortress, whose heavy defensive firepower caused problems. The Zero pilots switched to head-on attacks, and on 8 February they shot down two B-17s using this method.

In the meantime, the whole of the Tainan Group had moved to Balikpapan, where it was joined by the 3rd Air Group on 2 February. Both Groups were now well placed to conduct air superiority operations over eastern Java, and in the days that followed Allied aircraft fell in large numbers to the guns of the Zeros.

On 3 February, 27 Zeros of the 3rd Air Group and 27 of the Tainan Group engaged in a major air battle with American and Dutch air units over Soerabaya. The Allies lost 39 aircraft in air combat and a further 21 on the ground. Altogether, around 90 Allied aircraft were destroyed that day, giving the Japanese almost complete air superiority over Java.

Allied fighters encountered by the Zeros included RAF Hawker Hurricanes, evacuated to Java from Sumatra; Curtiss P-40s; Brewster Buffaloes; Curtiss Hawk 75As; and Curtiss-Wright CW-21Bs.

By 3 March 1942, the battle for Java was over, and on that day the 3rd Air Group, moving up to Kupang, deployed 17 Zero fighters under the command of Lt Miyano to attack the towns of Broome and Wyndham, in north-western Australia. The Zeros strafed a number of flying boats at their moorings and set them ablaze.

The 3rd Group had now been divided, part returning to Japan. The primary task of the remainder was to provide escorts for bombers attacking Darwin, where the main opposition consisted of the Curtiss P-40s of the USAAF's 49th Pursuit Group. In April, the Tainan Air Group deployed to Rabaul in New Britain, from where its main echelon of 24 Zeros moved up to Lae in eastern New Guinea for operations against the vital Allied base of Port Moresby. In the four-month period from April to July the Zeros claimed to have destroyed 201 Allied aircraft in air combat, with a further 45 probably destroyed, in the Port Moresby area alone, for the loss of only 20 Japanese fighters. The majority of the Allied fighters encountered during this period were Bell P-39 Airacobras, which were no match at all for the Zero. During the early weeks of 1942 Zeros also provided escorts for bombers attacking Singapore and Rangoon, the Burmese capital.

Returning to their home base to rest and replenish following the Pearl Harbor attack, the aircraft carriers of the 1st Air Fleet sortied again in January 1942, proceeding independently to support Japanese amphibious operations in the Pacific. *Akagi*'s air group attacked Rabaul and other targets in the Bismarck Archipelago, and in February participated in the air raids on Port Darwin, her Zeros shooting down four Allied aircraft on the 19th and destroying eight more on the ground. *Kaga* and *Soryu* also took part in these operations. *Hiryu*, meanwhile, had been detached to support the assault on Wake Island; her air group suffered some losses during this operation, and she returned to Japan to make good the damage. *Shokaku* and *Zuikaku*, their fighter element now increased to 21 Zeros each, also supported amphibious operations during January 1942.

Foray into the Indian Ocean

On 4 April 1942, a Catalina flying boat sighted a Japanese task force approaching Ceylon from the south and radioed its position minutes before it was shot down. The enemy force was Admiral Nagumo's 1st Carrier Striking Force, comprising the aircraft carriers *Akagi*, *Hiryu*, *Shokaku*, *Soryu* and *Zuikaku* – the ships whose aircraft had wrought such havoc on Pearl Harbor and Port Darwin – accompanied by four battleships, three cruisers and nine destroyers. Between them, the five carriers mustered some 300 strike aircraft and fighters; the Commander-in-Chief of the British Eastern Fleet, Admiral Somerville, had air assets consisting of 57 strike aircraft and 40 fighters on his carriers, with the Fulmars of Nos 803 and 806 Squadrons based ashore at Ratmalana. There were also 50-odd RAF Hurricanes delivered by HMS *Indomitable*.

The Japanese intention was clearly to destroy the Eastern Fleet. For the British this was a most dangerous move; for all they knew, the Japanese might try to seize the island, from where not only India but also the ocean supply routes to the Middle East could be threatened. This was at a time that all supplies had to be routed around the Cape; Germany's General Erwin Rommel was driving the Eighth Army back

An A6M2 Zero launching from the carrier *Zuikaku* early in 1942. (NASM)

A6M2 Zero fighters on the carrier *Akagi*. (NASM)

towards Egypt, and a new German offensive was building in Russia. The possibility of a German-Japanese link-up in the Middle East seems far-fetched with hindsight, but it was of very real concern to Churchill and his colleagues. Later, he was to call it the most dangerous moment of the war.

As soon as the news of the approaching Japanese task force was received, Admiral Layton, the Commander-in-Chief Ceylon, ordered every ship that could do so to sail from Colombo harbour. The cruisers *Cornwall* and *Dorsetshire*, which had been detached earlier on Somerville's instructions, were also ordered to rejoin Force A, the fast group of the Eastern Fleet which included the aircraft carriers and the battleship HMS *Warspite*. Early on 5 April – Easter Sunday – the Japanese fleet was sighted by a second Catalina, and soon afterwards it launched a strike of 53 Nakajima B5N high-level bombers and 38 Aichi D3A dive-bombers, escorted by 36 Zero fighters, to attack Colombo. Fierce air battles developed over the city and harbour as the raiding force was intercepted by 42 Hurricanes and Fulmars; seven Japanese aircraft were destroyed, but 19 British fighters were shot down. The attack caused heavy damage to built-up areas; the damage to shipping and the port installations was relatively light, although the auxiliary cruiser *Hector* and the destroyer *Tenedos* were sunk.

At about noon, the cruisers *Cornwall* and *Dorsetshire* were sighted by a reconnaissance aircraft from the heavy cruiser *Tone* and 53 D3A dive-bombers were immediately sent out to attack them. The bombing was devastatingly accurate and both ships were sunk, 1,112 men (of a total of 1,546) being rescued later by the cruiser HMS *Enterprise* and two destroyers. Albacores from the *Indomitable* later made a night radar search for the enemy force, but it had withdrawn to the south-east to refuel before heading back north to strike at Trincomalee naval base. At this time Admiral Somerville's Force A was steaming towards Ceylon from Addu Atoll, with his slow division (Force B) a long way behind; his ships were at times only 200 nautical miles from Nagumo's task force, but neither side made contact with the other. Addu Atoll had been set up as the Eastern Fleet's secret base, and Somerville, unable to locate the enemy, turned back towards it to safeguard it against a possible surprise attack.

Expected Attack

On 8 April, a Catalina again established contact with the Japanese carrier force 400 nm to the east of Ceylon and the ships at Trincomalee were ordered to put to sea. All units – including the light carrier *Hermes* – were able to get clear before the expected attack by 91 high-level dive-bombers, escorted by 38 fighters, developed early on the 9th. Of the 23 Hurricane and Fulmar fighters sent up to defend the harbour, nine were shot down, as were five out of a formation of nine Blenheim bombers sent out to try to locate the enemy force. Only light damage was inflicted on the target, but on the way back to their ships the Japanese aircrews sighted several ships, including the *Hermes,* the Australian destroyer *Vampire*, the corvette *Hollyhock* and two tankers. Three hours later, 80 dive-bombers arrived on the scene and sank all three warships and the tankers about 65 miles from Trincomalee. The *Hermes*, which had no aircraft on board, radioed desperately for help, but the surviving fighters at Trincomalee were in no position to offer it.

Meanwhile, part of Nagumo's force – a light carrier and six cruisers under Admiral Ozawa – had been detached to operate against shipping in the Bay of Bengal. In a five-day spree beginning on 7 April Ozawa destroyed 23 merchant ships totalling 112,312 tons, to which Japanese submarines operating off the west coast of India added a further 32,400 tons. Fortunately perhaps for the Eastern Fleet (whose Force A deployed to Bombay

The Japanese light carrier *Shoho* on fire and sinking after being attacked by the USS *Lexington*'s air group in the Battle of the Coral Sea. (US Navy)

immediately after the attacks on Ceylon, and Force B to East Africa to protect the convoy route) a Japanese task force never again made an appearance in the Indian Ocean.

Battle of The Coral Sea

During 4-8 May 1942, in the first naval engagement in history fought without opposing ships making contact with one another, United States carriers forces prevented a Japanese attempt to land an invasion force at Port Moresby, New Guinea, by turning back the enemy's covering carrier force, the latter comprising the aircraft carriers *Zuikaku*, *Shokaku* and *Shoho*. The opening phase of the battle began on 3 May, when Japanese troops landed on the island of Tulagi. Task Force 17 (Rear Admiral F.J. Fletcher) was despatched with the carrier USS *Yorktown* to attack the invasion force. On 4 May, the *Yorktown* launched a strike force comprising Douglas SBD Dauntless dive-bombers of Squadron VS-5, Douglas TBD Devastator torpedo-bombers of VT-5, and Grumman F-4F Wildcats of VF-5. In the attack, the Japanese destroyer *Kikutsuki* and a number of other vessels were destroyed, compelling the Japanese to evacuate their landing force.

On 5 May, the *Yorktown* joined other Allied naval units, including Task Force 11 (Rear Admiral A.W. Fitch) with the carrier USS *Lexington* south of the Louisiades, and after stationing an attack group in the probable track of the enemy transports, moved northward in search of the enemy covering force. On 7 May the light carrier *Shoho* was attacked and sunk by TBDs and SBDs of the *Lexington*'s air group, together with TBDs of VT-5; meanwhile, Japanese aircraft attacked the separately operating attack group, sinking the destroyer USS *Sims* and the fleet oiler USS *Neosho*. On 8 May, the fleet carrier *Shokaku* was attacked and heavily damaged by aircraft from both US carriers, but Japanese carrier aircraft, in a parallel attack, hit the *Lexington* with two torpedoes and damaged the *Yorktown* with a bomb hit below decks, although she remained operational. Later, fuel vapour in the *Lexington* exploded, causing uncontrollable fires; the carrier was abandoned and sunk by a US destroyer. Although a tactical victory for the Japanese, the Battle of the Coral Sea left them with no option but to call off the Port Moresby invasion, resulting in a clear strategic victory for the Allies.

Midway: The Decisive Battle

Six months after Pearl Harbor, a crushing humiliation was about to descend on the Japanese. At the beginning of June 1942, the Japanese launched a strong thrust in the central Pacific; its objective, as part of Yamamoto's eastward expansion plan, was to occupy Midway Island. The thrust was led by a four-carrier Mobile Force

Left: The US Navy suffered too. Here, the carrier USS *Lexington* is being abandoned by her crew after fuel explosions caused uncontrollable fires. (US Navy)

Below: A6M2-N floatplanes photographed by a US reconnaissance aircraft in Holtz Bay, Attu Island, in the Aleutians. (US National Archives)

comprising the *Akagi*, *Kaga*, *Hiryu* and *Soryu*, supported by heavy units of the First Fleet and covered by a diversionary attack by carrier aircraft on Dutch Harbor in the Aleutians.

The thrust towards Midway was met by a greatly outnumbered US carrier force composed of Rear Admiral Fletcher's Task Force 17, with the USS *Yorktown*, and Rear Admiral R.A. Spruance's Task Force 16, with the USS *Hornet* and USS *Enterprise*, supported by Navy, Marine Corps and Army air units based on Midway.

The opening action of the battle took place on 3 June when the Japanese fleet was sighted by a PBY Catalina; six Grumman TBF Avengers of Torpedo Squadron VT-8's shore-based detachment were launched to attack the Japanese, together with Army Air Force B-17s and Marine Vought SB2U Vindicators. None of the Avengers scored a hit and only one returned to Midway. At 0700 on 4 June the *Enterprise* and *Hornet* launched their strike groups, comprising 14 TBDs of VT-6 and 15 of VT-8, with F4Fs of VF-6 flying top cover. VT-8 attacked first and all 15 aircraft were shot down by Zeros; only one crew member survived. VT-6, attacking the *Kaga* in turn, lost ten aircraft before they reached their dropping points. The *Enterprise*'s air group, attacking the *Soryu* with 12 TBDs of VT-3 and 17 SBDs of VB-3, escorted by six F4Fs of VF-3, fared no better. Only five TBDs survived to make their torpedo attacks, and three of these were shot down on the way out. Of the 41 TBDs launched, only six returned to the task force, and one of these ran out of fuel and ditched.

The sacrifice of the three VT squadrons was not in vain. They had absorbed the bulk of the enemy fighter attacks, and the Zeros were still scattered when 37 Dauntless dive-bombers from the *Enterprise*'s VB-5 and the 17 from *Yorktown*'s VB-3 made their attack, sinking the *Akagi*, *Kaga* and *Soryu*. The cost to the dive-bombers was 16 aircraft lost from *Enterprise*'s air group. A Japanese counter-attack from the *Hiryu* damaged the *Yorktown*, but she returned to full operation after a short time. Then a second attack was made by six B5N *Kate* torpedo-bombers; two were shot down, but the other four launched their torpedoes and two hit the carrier, which had to be abandoned. She was later sunk by a submarine. At 1700, the *Hiryu* was crippled in an attack by 24 SBDs from the *Enterprise*; her burned-out hulk was sunk by a Japanese destroyer the next day. For the cost of 92 aircraft and the *Yorktown*, the US Navy had destroyed four fleet carriers, three-quarters of the Imperial Japanese Navy's carrier strike force. With control of the air irretrievably lost, the Japanese withdrew on 5 June under attack by Midway-based and carrier aircraft; on 6 June, the cruiser *Mikuma* was sunk and her sister ship *Mogami* severely damaged in an air strike from the USS *Enterprise*.

Other US losses in the Midway battle were 40 shore-based aircraft and the destroyer *Hammann*, sunk by the same submarine that torpedoed the crippled *Yorktown*. In addition to the four carriers, the Japanese lost 258 aircraft and a large percentage of their most experienced carrier pilots. Midway was a decisive defeat that put an end to Japan's offensive plans, and effectively turned the tide of the Pacific war.

Wildcat v Zero

In the months that followed the Japanese attack on Pearl Harbor in December 1941, the fighter aircraft that really held the line for the United States was the Grumman F4F Wildcat. Powered by a Pratt & Whitney Twin Wasp radial engine and armed with four .50-calibre machine guns, the tubby little Wildcat was the standard single-seat

A US air strike hits Japanese shipping off Rabaul, February 1942. The warship at the forefront is a Myoko-class cruiser. (USAF)

fighter in service with the US Navy and US Marine Corps at that time. Although completely outclassed by the Mitsubishi A6M Zero, it fought valiantly against hopeless odds in those early months of the Pacific war, and many American aces scored their first victories while flying it.

The Wildcats first met the Japanese in action during the heroic defence of Wake Island in December 1941, shooting down nine Japanese aircraft before they were overwhelmed. The first kill by a carrier-based Wildcat in the Pacific was scored by Lieutenant (Jg) W.E. Rawie, who was strafing Japanese positions during an attack on the Japanese-held Marshall and Gilbert islands by the aircraft carriers USS *Yorktown* and USS *Enterprise* on 1 February 1942 when he was attacked by two Mitsubishi A5M Claude fighters. He turned into them and shot one of them down as they went past.

A second US carrier strike was laid on later in the month, this time by the USS *Lexington* against Rabaul, the idea being to try to relieve growing Japanese pressure on Port Moresby in New Guinea. Part of *Lexington*'s carrier air group was VF-3, a Wildcat squadron commanded by Lieutenant Commander John S. Thach. An excellent shot and the US Navy's leading exponent of air tactics, Thach had developed a defensive manoeuvre which became known as the Thach Weave. This involved elements of two aircraft, with about 300 yards' spacing between each aircraft so that each pilot could watch his partner's tail. If a pilot was attacked from astern, he would break towards his partner, who would also break outwards towards the second pair. This pair would then break inwards in order to meet the threat head-on; the attacking aircraft would effectively be drawn into a trap. It proved an excellent tactical manoeuvre, and with modification it is still used today.

Thach had insisted on his pilots constantly striving to improve their marksmanship, so that in February 1942 VF-3's gunnery record was probably the best in the US Navy. Their skill was to stand them in good stead, and compensated in part for the Wildcat's shortcomings. It was also appropriate that Thach himself had the privilege of scoring VF-3's first kill, a Kawanishi H6K reconnaissance flying boat which he shot down in flames 400 miles north-east of Rabaul. The Japanese crew made no attempt to abandon their aircraft and the gunners kept on firing until the Mavis exploded on the sea.

It was soon clear that the Japanese had remained at their posts long enough to alert their HQ on Rabaul to the presence of an American aircraft carrier. Late that afternoon, nine Mitsubishi G4M bombers were seen approaching the USS *Lexington*. They were engaged by the carrier's combat air patrol (CAP) and the Wildcats shot down four of them, but the rest continued on course and bombed in formation from 11,000 feet. The bombing was accurate, but the carrier took evasive action and escaped damage. Then, as the Wildcats were preparing to land-on in order to refuel and re-arm, a second formation of nine G4Ms was detected by the carrier's radar.

Only two Wildcats were in a position to intercept this second attack, and their pilots showed no hesitation. In the leading fighter, Lieutenant H. 'Butch' O'Hare ripped into the Japanese formation and sent down two G4Ms in flames on his first pass. The G4M had an excellent range, but this had been achieved by saving weight at the expense of armour protection for the crew and fuel tanks; as a result, when the G4M was subjected to a determined attack it usually exploded into flames. O'Hare braved heavy defensive fire to destroy another bomber, quickly followed by two more. In just five minutes, he had become the US Navy's first air ace.

By this time, John Thach and other Wildcat pilots had arrived to join the fight, and they followed the surviving bombers for as long as their fuel permitted. Of the 18 Japanese bombers that tried to attack the *Lexington* that day, 13 went blazing into the sea. For his personal exploit, Butch O'Hare was awarded the Medal of Honor. He went on to score a total of 12 kills before being killed in action during a strike on the Gilbert Islands on 26 November 1943.

Later in the war, John Thach was assigned to the Fast Carrier Task Force as Air Operations Officer, where he developed tactics that played a major part in destroying Japan's offensive air capability. He continued his distinguished career after the war, eventually reaching the rank of rear admiral. He died on 15 April 1981.

Guadalcanal

On 7 August 1942, a division of United States Marines stormed ashore on Guadalcanal, in the Solomon Islands. For the Americans, it was the first step on a long journey back across the Pacific, a journey that would end almost exactly three years later with the nuclear destruction of Hiroshima and Nagasaki. One of the main American objectives in the invasion of Guadalcanal was to capture an airstrip which had been built there by the Japanese. The Marines went in and took it, and the land battle subsequently centred on the struggle to retain it. In one of the most tenacious and heroic actions of the Pacific war the Marines clung grimly to the perimeter they had set up around the strip, which they called Henderson Field in honour of Major Lofton R. Henderson, a squadron commander and Marine Corps pilot lost in the Battle of Midway, and by 20 August it had been made sufficiently secure for the first US fighters to fly in.

They were the Wildcats of Marine Fighter Squadron VMF-223, led by Major John L. Smith, and they were followed by Major

An aerial view of Henderson Field in late August 1944, with F4F Wildcat fighters visible. (US Navy)

US Marine Corps aviation aces pose with an F4F Wildcat on Gualdalcanal. Medal of Honor recipients John Smith and Robert E. Galer are on the left and right; the pilots in the middle are Ken Fraizer and John Dobbin. (US Navy)

Robert E. Galer's VMF-224 a few days later. The day after their arrival on Guadalcanal, the pilots of VMF-223 intercepted six Zeros at 14,000 feet over the island, and Major Smith shot down one of them to score the squadron's first victory. The following afternoon the Japanese came again, this time with 15 bombers escorted by 12 Zeros. All of VMF-223's serviceable Wildcats took off to intercept the enemy, and in the course of a savage air battle they destroyed 16 Japanese aircraft for the loss of three Wildcats. Joe Smith and one of his flight commanders, Captain Marion E. Carl, each shot down three.

Day after day, while the ground forces strove desperately to hold the thin perimeter around Henderson Field, the Marine pilots went into action against the Japanese squadrons that were making determined efforts to wipe out the primitive airstrip. At the same time, Japanese warships shelled the base every night, and individual enemy aircraft carried out nuisance raids to ensure that the American personnel got little rest. As the weeks went by, malaria, dysentery and fatigue began to have a telling effect, yet the Americans, flying to the limits of their physical endurance, somehow managed to retain air superiority.

By the time VMF-223 was relieved in October 1942, the pilots had destroyed 110 enemy aircraft. John Smith's score was 19, while his close rival Marion Carl had shot down sixteen. Major Robert E. Galer, of VMF-224, had chalked up 13 victories; both he and Smith were awarded the Medal of Honor. All three survived the war.

The replacement squadrons of Guadalcanal were VMF-121 and VMF-212. One of the former's pilots was Captain Joe Foss, a farm boy from South Dakota whose marksmanship, thanks to his father's tuition with rifle and shotgun, was superb. Foss rose to fame with incredible speed over Guadalcanal; by the middle of October he was averaging one victory a day, and by the end of the month three a day. On 23 and 25 October he destroyed a total of nine enemy aircraft, all of them Zeros. By the time VMF-121 left Guadalcanal in January 1943, its pilots had destroyed 123 Japanese aircraft for the loss of 14 Wildcats.

Joe Foss's personal score was 26, which made him the first American to equal the personal score of Eddie Rickenbacker, the leading American ace of the Great War. His outstanding combat record earned Joe Foss the Medal of Honor. He never flew in combat again – American policy being that outstanding combat pilots who had been awarded the Medal of Honor should not be required to risk their lives in action again – and, after the war, he became Governor of South Dakota. Brigadier General Joseph J. Foss died in hospital on 1 January 2003 at the age of 87, having suffered an apparent aneurism.

The exploits of Joe Foss tended to overshadow all others in the embattled sky over Guadalcanal; but in the hectic days of October and November 1942, when the Japanese were making their most determined attempts to recapture the island, the effective resistance of the American pilots was due in no small measure to the dedicated leadership of one man. He was Lieutenant Colonel Harold W. Bauer, known throughout US Marine Aviation as 'Indian Joe'.

Bauer was the commanding officer of VMF-212, which was based on Hefate in the New Hebrides when the battle for Guadalcanal began. His initial task was to provide a pool of trained replacement Wildcat pilots for the two Marine squadrons already in action on the island, and before long he was making frequent visits to Guadalcanal himself, flying combat air patrols with the others. On 28 September, he destroyed an Aichi D3A dive-bomber, his first victory, and on 3 October he claimed four Zeros, part of a Japanese fighter force which was escorting a large group of G4M bombers. He shot down four more enemy aircraft – all D3As – on 16 October, the day that he led his own VMF-212 into action on the island.

It was Bauer who instructed his Wildcat pilots to dogfight with the Zeros. This had been attempted earlier, with disastrous results because the Zero was more manoeuvrable than the Wildcat, but Bauer had shrewdly assessed that the Japanese must now be fielding less experienced pilots. He was right, and in a big air battle on 23 October, 24 Wildcats and four Army P-39s took on a big formation of G4Ms and Zeros and shot down 20 fighters and two bombers. All the enemy aircraft fell in the vicinity of the airfield in clear view of the exhausted defenders, providing a massive boost to their morale.

On that day, Bauer was appointed Commander of Fighters on Guadalcanal. On 14 November, after shooting down another Zero – his eleventh victim – while providing top cover for US bombers which were attacking Japanese shipping off the island, he was himself hit and sent down into the sea. The other pilots saw him in his lifejacket, waving, and sent out a Grumman Duck amphibian to pick him up, but despite an extensive air and sea search Indian Joe Bauer was never found. He was posthumously awarded the Medal of Honor.

Battle of Santa Cruz

In October 1942, the Japanese launched a determined campaign to capture and neutralize Henderson Field on Guadalcanal as a preliminary to the destruction of the remaining US naval air resources in the Solomons. At this point the US Navy had only two carriers in the area, the USS *Enterprise* and the USS *Hornet*, the latter having replaced the USS *Saratoga*, damaged by a submarine torpedo on 31

Captain Joe Foss (left) and Wildcat pilots on Guadalcanal, early 1943. (via World War Photos)

A6M Zeros preparing to launch from the carrier *Shokaku* during the Battle of Santa Cruz, October 1942.

August and withdrawn to Pearl Harbor for repair. The Americans had also lost the USS *Wasp*, sunk by a submarine on 15 September while escorting a troop convoy to Guadalcanal.

To activate their plan, the Japanese naval forces put to sea from Truk on 11 October. They included the Carrier Striking Force with the *Shokaku* and *Zuikaku*, the light carrier *Zuiho* and seven destroyers; the Advance Striking Force with the new carriers *Junyo* and *Hiyo* supported by two battleships, five cruisers and 13 destroyers; and the Battleship Striking Force with two battleships, three cruisers and eight destroyers. The latter subjected Henderson Field to a very heavy bombardment on 13 October and on three successive nights; aircraft losses were partly made good by the deployment of 19 F4Fs from Espiritu Santo, together with seven SBDs. The total number of aircraft available to defend Henderson Field against attacks by enemy forces (which landed on 15 October) was now sixty.

On 26 October 1942, the opposing carrier task forces located one another almost simultaneously and launched their strike

Top: A6M3 Zeros of the *Zuikaku* Fighter Squadron seen on Rabaul. (via John McKenna)

Above: The battleship USS *South Dakota* comes under attack by Mitsubishi B5N2 torpedo-bombers during the Battle of Santa Cruz. (US Navy)

aircraft. The Japanese force, in two waves 45 minutes apart, comprised 42 Aichi D3A2 dive-bombers, 36 Mitsubishi B5N2 torpedo-bombers and 55 Zero fighters from the *Shokaku*, *Zuikaku* and *Zuiho*; the *Hornet* and *Enterprise* launched three waves comprising 30 SBD Dauntlesses, 20 TBM Avengers and 24 F4F Wildcats. While these forces were en route, two SBDs from the *Enterprise*, on an armed reconnaissance, encountered the *Zuiho* and bombed her, damaging her flight deck and rendering it unusable.

At 0940, purely by chance, the opposing air groups ran into each other and a brief air battle developed in which the Zeros shot down three SBDs and four Wildcats for the loss of five of their own number. At 1010 the first Japanese wave located the *Hornet* and subjected her to a bomb and torpedo attack that left her burning and listing in the water. Thirty minutes later, the *Hornet*'s strike aircraft also found the *Shokaku* and scored four hits with 450kg (1000lb) bombs, putting the carrier out of the battle. The other US strike aircraft located the enemy cruiser and destroyer forces and attacked with bombs and torpedoes, damaging one cruiser. Meanwhile, the second Japanese wave attacked the *Enterprise*, scoring three bomb hits; torpedo attacks were frustrated by the Wildcat combat air patrols. Further attacks in the afternoon, launched by the *Zuikaku* and *Junyo*, succeeded in sinking the crippled *Hornet*. Fortunately for the Americans, the Japanese were unable to exploit their tactical successes. They had exhausted their fuel reserves and were compelled to withdraw to Truk to replenish.

The battle for the Solomons showed that the Zero, while still superior to the American types it encountered, no longer enjoyed the complete mastery of the air that had enabled it to dominate operations in the early months of the Pacific war. During the lengthy battle for Guadalcanal, the Zero units faced a further problem in that they were forced to operate from land bases located some 650 miles from the combat area. Zero losses mounted alarmingly, as in addition to combat losses sustained from the Japanese attempt to counter the first American amphibious landings on the island, a large number of A6M3 Model 52s were lost when they ran out of fuel during the long over-water flight home. The Sakai 21 engine was found to have a higher fuel consumption that the Sakai 12 of the A6M2, and the tank capacity had been reduced when the engine was installed. At the request of operational units Mitsubishi fitted a 9.9 Imp gal (45l) tank in each wing outboard of the cannon bay, and in order to retain a low wing loading, in spite of the increased weight, the folding wingtips were again fitted. The installation of the wing tanks restored the range of the Zero to its former level, but any advantage would soon be eliminated by the appearance of new and formidable American fighters in the Pacific theatre.

New Fighters, New Tactics

The first was the Lockheed P-38 Lightning, which deployed to Guadalcanal with detachments of the 347th Fighter Group from New Caledonia from late 1942. On 18 April 1943, P-38 pilots of the 339th Fighter Squadron, 347th Fighter Group, pulled off a spectacular coup when they flew to the limit of their combat range from Guadalcanal to Bougainville and shot down a G4M bomber carrying Admiral Isoroku Yamamoto, Commander-in-Chief of the Japanese Combined Fleet and architect of the attack on Pearl Harbor. None of the six Zeros escorting Yamamoto's G4M (and another carrying members of his staff) was lost, but one P-38 was shot down into the sea.

Admiral Isoroku Yamamoto, Commander-in-Chief of the Japanese Combined Fleet and architect of the attack on Pearl Harbor.

By the end of 1942, a system of code names was being used by the Allies to identify Japanese aircraft. Fighters were allocated boys' names, and the A6M2 and A6M3 received the name Zeke. The later A6M5 was originally allocated the name Hap, which was the nickname of General Henry H. Arnold, commanding the US Army Air Forces, but Arnold was not pleased with the choice so it was changed to Hamp. In practice, except perhaps in official reports, the name Zero continued to be universally used by the Allies.

Early in 1943 some US Marine Corps squadrons began to re-equip with a powerful new fighter, the Vought F4U Corsair, which deployed to Guadalcanal in February. One pilot in particular scored spectacular successes while flying this type; he was Lieutenant Robert Hanson of Marine Squadron VMF-215, and he rose to fame in the bitterly contested sky over Rabaul. On 14 January 1944, Hanson fought the first of a series of combats that was to set a record, destroying five out of a formation of 70 Zeros that were trying to intercept American bombers. His next five sorties over Rabaul netted him 15 more enemy fighters, bringing his score to 20 enemy aircraft destroyed over a period of only 17 days.

The US fighter that really changed the course of the Pacific air war, however, was the Grumman F6F Hellcat. The Hellcat flew for the first time on 26 June 1942, its design having benefited from combat lessons learned by its predecessor, the Wildcat. First deliveries of the Grumman F6F-3 Hellcat were made on 16 January 1943, and the aircraft saw its first combat over Marcus, one of the Caroline Islands, on 31 August. The Hellcat went on to play a prominent role in all US naval operations.

During 1943 there was a marked and noticeable decline in the level of expertise of Japanese pilots. The loss of the nucleus of the Imperial Japanese Navy's best pilots, many of whom had seen extensive combat in China before the start of the Pacific war, in the great battles of 1942 was at last beginning to have its effect. As new American fighter types arrived in the Pacific, the Japanese began to suffer combat losses that were little short of staggering.

Spitfires in Action

The Supermarine Spitfire also played its part in the Pacific air war, forming a component of the air defences of northern Australia. In 1942, after Japanese carrier aircraft had first attacked Darwin, three squadrons of Spitfires had been shipped to Australia at the urgent request of the Australian government. The three squadrons – Nos 452 and 457 RAAF and No. 54 RAF – were formed into No. 1 Fighter Wing, RAAF, under the command of Squadron Leader Clive Caldwell, a highly skilled and experienced fighter pilot who had already regsitered 20 kills in the Middle East.

The Spitfire variant that was sent to Australia was the 'tropicalized' Mk Vc. In a turning fight, and in the climb, it was outclassed by the Zero fighter, as the pilots

Oil tanks ablaze at Darwin after an attack by Japanese carrier aircraft in 1942. (Australian War Memorial)

Spitfire pilot Flying Officer Ian Taylor of No. 54 Squadron RAF walks away from his Mk Vc (Trop). (Australian War Memorial)

Wreckage of an A6M2 Zero of the Tainan Air Group, shot down over Port Moresby. (Australian War Memorial)

of No. 1 Fighter Wing were soon to learn. In February 1943, the Japanese re-opened their bombing offensive against northern Australia, and on the 6th of that month Flight Lieutenant R.W. Foster opened the Spifire's scoreboard in the theatre by shooting down a Mitsubishi Ki.46 reconnaissance aircraft 35 miles off Cape van Diemen.

On 2 March, Caldwell, now a wing commander, destroyed a Zero and a Nakajima B5N torpedo-bomber, while Squadron Leader A. Thorold-Smith, commanding No. 452 Squadron, also shot down a Zero. The latter pilot was killed on 15 March, when the wing intercepted 14 Japanese aircraft over Darwin and claimed seven of the enemy for the loss of four Spitfires.

The first major battle occurred on 2 May, when the Japanese sent in a force of 18 bombers and 27 Zeros from Timor. The Japanese were detected by radar while they were still a long way out to sea – 49 minutes' flying time from the coast, in fact – and the wing's 33 Spitfires were all airborne within 15 minutes, climbing hard to meet the raiders. When the Spitfires reached 26,000 feet, however, Caldwell saw that the Japanese formation was still about 4,000 feet higher up. To attack it on the climb would have been foolhardy, for the nimble Zeros would have held all the advantages, so Caldwell delayed while his Spitfires got into position above the enemy, with the glare of the sun behind them. This meant that the Japanese were able to bomb Darwin without meeting any fighter opposition, a fact that later caused a huge outcry in the popular press, but Caldwell was quite right in his decision.

Caldwell's Spitfires, unseen against the sun, shadowed the Japanese until they were out over the Timor Sea, then he ordered No. 54 Squadron to attack the Zeros while the other two squadrons took on the bombers. The Spitfires went into the attack almost vertically and a furious air battle developed as the Zero pilots, recovering from their surprise, turned to meet the attackers. When the battle was over, five Zeros had been destroyed, but on the debit side five Spitfires had been shot down, two of the pilots being killed, and five more had been compelled to make forced landings after running out of fuel.

In the weeks that followed, however, Caldwell continued to develop the wing's tactics, ignoring the growing storm of criticism about the celebrated Spitfire's lack of success, and one day in July 1943 his efforts paid dividends. On that day, No. 54 Squadron was scrambled to intercept a raid on Darwin by 47 bombers and their fighter escorts. Only seven Spitfires reached the attackers, but they shot down seven bombers and two Zeros for no loss.

Then, on 20 August, three Japanese reconnaissance aircraft appeared over Darwin, heralding another raid; the

Spitfires shot down all three of them. The Japanese sent another; it was shot down by Clive Caldwell, his 28th and last victory. The Japanese sent yet another, this time under strong fighter escort. No. 54 Squadron was scrambled to intercept, and the Zeros fell on the Spitfires as they climbed, shooting down three of them. But the Spitfires in turn destroyed one Zero and damaged two more so badly that it is almost certain they came down somewhere in the Timor Sea.

Soon afterwards, the Japanese daylight raids on Darwin ceased and the enemy switched to sporadic night attacks which were to continue, with little effect, until early in 1944. Caldwell's Spitfires had achieved their objective. After the war, Clive Caldwell, who retired as a group captain, became a partner in a cloth importing business in Sydney. He died on 5 August 1994.

In Burma, RAF Spitfires began to encounter increasing numbers of A6M3 Zero Model 32 aircraft from the beginning of 1944. These aircraft served alongside the Nakajima Ki.43 Hayabusa (Oscar) in mixed fighter units. In the closing months of the war, the Spitfire's naval counterpart, the Seafire, also saw action against the Zero while serving aboard the carriers of the British Pacific Fleet, although the Seafires operated mainly in the fleet defence role and the Zeros they engaged were mostly kamikazes.

By the end of 1943 the Imperial Japanese forces had been forced on to the defensive. The loss of the Imperial Japanese Navy's major aircraft carriers meant that their fighter squadrons mostly operated from shore bases, which came under repeated attack not only from aircraft such as the long-range B-24 Liberator, but also from medium bombers like the B-25 Mitchell and B-26 Marauder, now able to operate from newly captured island airfields. Enemy shipping, attempting to reinforce and resupply the Japanese island outposts, also came under constant air attack.

Turkey Shoot

The most devastating loss suffered in the air by the Japanese during the entire Pacific war occurred during the Battle of the Philippine Sea in June 1944, when carrier-based aircraft provided air cover for the occupation of the Marianas. During the first major fighter sweep over the islands, on 11 June, the carrier aircraft destroyed one-third of the defending air force. On the 19th, with the amphibious invasion in full swing, large numbers of Japanese bombers and torpedo-bombers made a series of desperate attempts to hit the task force; they were detected by radar at a range of 150 miles (240km), and the carrier fighters were waiting for them. The great air battle that followed was a one-sided massacre that would go down in history as the 'Marianas Turkey Shoot'. American combat air patrols and anti-aircraft fire destroyed 325 enemy aircraft, including 220 of the 328 launched by Japanese aircraft carriers. American losses were 16 Hellcats in combat, and seven other aircraft shot down by Japanese fighters or ground fire.

The Japanese aircraft carrier *Zuikaku* and two destroyers under attack by US naval aircraft on 20 June 1944. (US Navy)

Japanese shipping under attack, Rabaul, November 1943. One of the attacking B-25s can be seen at lower left, flying at extremely low level. (National Archives)

Pacific War: Final Phase October 1944–August 1945

Leyte Gulf

In October 1944, the reconquest of the Philippines began with the American landings at Leyte. On learning of the invasion, the C-in-C Japanese Combined Fleet, Admiral Toyoda, ordered Vice-Admiral Kurita's Centre Force to sail from Brunei in Borneo on 22 October with the battleships *Yamato, Musashi, Nagato, Kongo* and *Haruna*, two cruisers and 15 destroyers. This was followed by Vice-Admiral Nishimura's Southern Force, with the battleships *Fuso* and *Yamashiro*, one cruiser and four destroyers.

On 24 October, the Centre Force was attacked by four waves of American aircraft from the Fast Carrier Force. During these attacks the *Musashi* was hit by ten bombs and six torpedoes and sank in about eight hours, with the loss of 1,039 of her crew. The Yamato was also hit by two bombs but these had little effect on her. Japanese air attacks resulted in the loss of the light carrier USS *Princeton*. In the Southern Force, the *Fuso* was sunk by gunfire and torpedoes in Surigao Strait. This force was then engaged by the battleships *West Virginia, California, Tennessee, Maryland* and *Mississippi*, as well as US and Australian cruisers. Vice-Admiral Nishimura's flagship, the *Yamashiro*, was sunk, hit by numerous shells and three torpedoes; her commander went down with her.

The Japanese were to suffer still more losses. At dawn on 25 October, Task Force 34 was formed from the US battleships *Iowa, New Jersey, Washington, Alabama, Massachusetts* and *Indiana*, four cruisers and ten destroyers with the objective of destroying a diversionary force, approaching from the north under Vice-Admiral Ozawa. Instead, Admiral Halsey, flying his flag on *New Jersey*, took his ships south to hunt down what remained of Kurita's force, leaving carrier aircraft to deal with Ozawa. This tactic proved successful, and the carriers *Chitose, Zuikaku, Zuiho* and *Chiyoda*, as well as a destroyer, were all sunk.

An A6M5 Zero armed with anti-personnel bombs, seen in the Philippines. (USAF)

Kamikazes

The Battle of Leyte Gulf spelled ruin for the Imperial Japanese Navy. Its carriers were gone, and although it still had surface warships, most were confined to

American soldiers inspecting an abandoned A6M5 Zero at an airstrip in the Marshall Islands.

harbour through lack of fuel. Losses at sea continued, too. On 21 November the battleship *Kongo* was torpedoed and sunk by the US submarine *Sealion* northwest of Keelung. Even so, the Allies had a clear indication that the assault on the last Pacific islands before Japan itself would be hard and costly, for the Battle of Leyte witnessed the first planned attacks by Japanese Kamikaze (Divine Wind) suicide pilots. In October 1944, four 'special attack' units were operational in the Philippines, each equipped with 12 Zeros armed with a pair of 227kg (500lb) bombs. The suicide pilots' primary targets were the aircraft carriers, and off Leyte they sank the escort carrier *St Lo* and damaged the *Sangamon*, *Suwannee*, *Santee*, *White Plains*, *Kalinin Bay* and *Kitkun Bay*. Further kamikaze attacks damaged the carriers *Intrepid*, *Franklin* and *Belleau Wood*; the *Intrepid* was hit again in November, and the *Lexington*, *Essex* and *Cabot* were also damaged.

The kamikazes were also in evidence during the American invasion and occupation of Luzon, the largest island in the Philippines, in January 1945, sinking the carrier *Ommaney Bay* and damaging the *Manila Bay*, *Savo Island*, *Kadashan Bay* and *Salamaua*. Further kamikaze attacks during the landings on Iwo Jima in February sank the escort carrier *Bismarck Sea* and seriously damaged the *Saratoga*, while in March, during operations off Okinawa, kamikazes damaged the US carriers *Enterprise*, *Intrepid*, *Yorktown*, *Franklin* and *Wasp*.

Kamikaze attacks on Allied warships persisted throughout April and May; during the period March–June 1945, 1,465 kamikaze aircraft took part in day and night attacks, claiming 26 ships sunk and 164 damaged.

The majority of the kamikazes were Zeros, their pilots sacrificing themselves in a bid to achieve some sort of victory in a war that was already lost.

The pilot of an A6M5 Zero (Zeke 52) runs up his engine. (via John McKenna)

This heavily camouflaged A6M5 Zero was found by US forces on Saipan, 1944. (NASM)

Uncovering the Zero's secrets

Although nine A6M2 Zeros were shot down during the attack on Pearl Harbor, the wreckage revealed little more than that the aircraft lacked armour and self-sealing fuel tanks. Technical information was

Australian soldiers inspecting an abandoned A6M3 Zero on New Guinea. (Australian War Memorial)

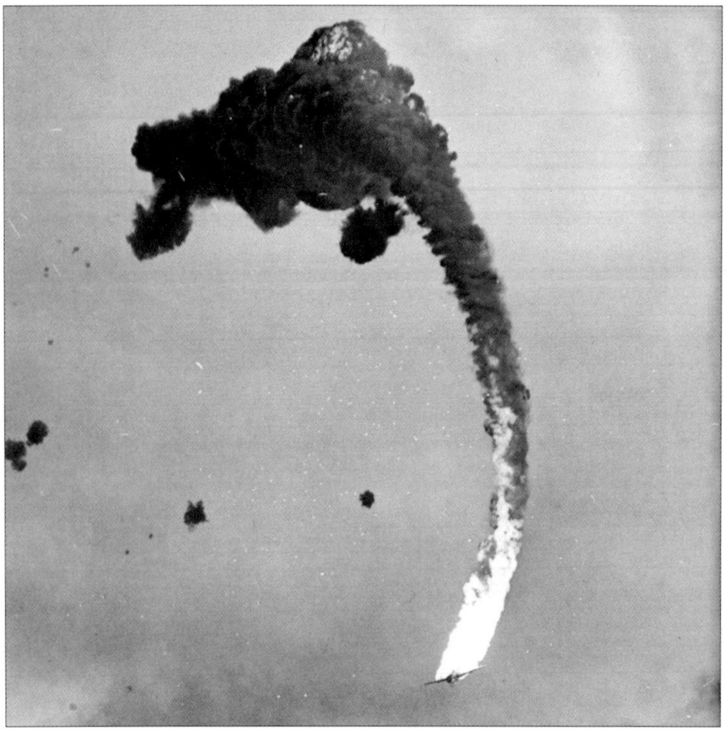

A kamikaze, in this case a Yokosuka D4Y *Judy*, meets a fiery end after being hit by anti-aircraft fire during an attack on US shipping off Okinawa. (US Navy)

This well-known image shows a Zero pilot desperately manoeuvring his aircraft in an attempt to crash into the battleship USS *Missouri*. He struck the hull side below the main deck, causing minor damage, but no casualties. (US National Archives)

The aircraft carrier USS *Bunker Hill* pictured seconds after being hit by two kamikazes off Okinawa. (US Navy)

also gleaned from three Zeros brought down elsewhere. One Zero, from the carrier *Hiryu* and flown by CPO Hajime Toyoshima, crash-landed on Melville Island during the bombing of Darwin on 19 February 1942. The Zero was badly damaged, although Toyoshima suffered only minor injuries and was taken prisoner. Another Zero came down near Cape Rodney, New Guinea, and might have been restored to airworthy condition had not the recovery team chopped off its wings, severing the main spar. A third Zero came from China, where a German-American engineer called Gerhard Neumann assembled a working A6M2 from parts salvaged from other crashed aircraft. The aircraft was eventually shipped to the USA, but in the meantime the Americans had received a real windfall in the shape of a Zero that was recovered intact.

In June 1942, as part of the Midway operation, the Japanese attacked the Aleutian Islands, off the south coast of Alaska, sending carrier aircraft to bomb Dutch Harbor on Unalaska Island. During the second raid, on 4 June 1942, a Zero from the carrier *Ryujo* flown by PO Tadayoshi Koga was brought down by small-arms fire. Koga attempted a forced landing on Akutan Island, 25 miles east of Dutch Harbor, but the ground was muddy and waterlogged and the fighter overturned, killing its pilot.

The crash site remained undetected and undisturbed for over a month before it was sighted by the crew of a PBY Catalina. A recovery team extricated the unfortunate pilot from the aircraft, which was assessed as repairable, and on 15 July the Zero was pulled from the mud and hauled overland to a barge, which took it to Dutch Harbor. After a preliminary clean-up the Zero was taken to Seattle on board a transport ship, the USS *St Mihiel*, and from there it was transported by barge to Naval Air Station North Island near San Diego, where repairs were carried out. The aircraft was pronounced airworthy on 20 September, and between that date and 15 October it made 24 test flights, the pilot being Lieutenant Commander Eddie R. Sanders. An extract from his report reads:

"These flights covered performance tests such as we do on planes undergoing Navy tests. The very first flight exposed weaknesses of the Zero which our pilots could exploit with proper tactics ... immediately apparent was the fact that the ailerons froze at speeds above 200 knots so that rolling maneuvers at those speeds were slow and required much force on the control stick. It rolled to the left much easier than to the right. Also, its engine cut out under negative acceleration due to its float-type carburettor. We now had the answer for our pilots who were being outmaneuvered and unable to escape a pursuing Zero: Go into a vertical power dive, using negative acceleration if possible to open the range while the Zero's engine was stopped by the acceleration. At about 200 knots,

Above: The Akutan Zero pictured during a test flight in the USA, September 1942. Although somewhat cramped by western standards, the 'birdcage' cockpit provided the pilot with a good all-round view. (via Philip Jarrett)

Left: The Akutan Zero, seen in US insignia at Langley, 1943. (US Navy)

US Navy personnel preparing the Akutan Zero for a test flight. (US Navy)

roll hard right before the Zero pilot could get his sights lined up."

The Akutan Zero was destroyed in a training accident in 1945, when a Curtiss SB2C Helldiver lost control and collided with it while both aircraft were taxiing. Parts of it are preserved in several US museums.

Non-Japanese Zero Operators

The Republic of China Air Force on Formosa (Taiwan) operated a small number of captured A6Ms after the war. The aircraft had originally been flown as part of Japan's 12th, 13th, 14th and 15th Naval Units as well as the Shanghai Kokutai

US Navy test pilot Lt Cdr Eddie R. Sanders taxies the Akutan Zero prior to making a test flight.

A captured A6M5 Zero pictured during a test flight in 1944. (NASM)

A fine shot of a captured A6M2 Zero under test in the United States, September 1944. (US Navy)

Left and centre: A captured A6M2, October 1943, seen at Wright Field near Dayton, Ohio, during its evaluation programme. (Harry Holmes)

A6M5 Zero fighters found abandoned under camouflage netting on Saipan, 1944.

on the mainland. They were also flown as part of Japan's Formosan Navy units, Hao Toko, Takao, Tetshu, Kagi, Toki and Tainan Kokutai. A6Ms were used by both sides, Nationalist and Communist, during the Chinese Civil War.

In 1945, Indonesian pro-independence guerrillas captured a small number of A6M aircraft at numerous Japanese air bases, including Bugis Air Base in Malang, re-occupied on 18 September 1945. Most aircraft were destroyed in fighting between the newly proclaimed Republic of Indonesia and the Netherlands during the Indonesian National Revolution of 1945–1949. The Royal Thai Air Force was supplied with various Japanese aircraft types between 1941 and 1945, and there is some evidence that Zeros were among these, although the main fighter type supplied was the Nakajima Ki-43 Hayabusa, 24 examples being delivered.

An A6M5 Zero in the markings of the Chinese Nationalist Air Force. (NASM)

The Zero Aces

In the European and American air forces, the tern 'ace' is applied to a fighter pilot who has shot down five or more enemy aircraft. Japanese records, compiled after the war, put the starting figure at eight, and 149 Zero pilots of the Japanese Navy come within this category. Post-war research has identified the following pilots as the top ten, although the list is open to doubt, as official sources and personal accounts differ widely:

- Tetsuzo Iwamoto: 94, including 14 in China (202 kills claimed in his personal diary). Survived the war, but died of sepsis after an operation in 1955.
- Shoitch Sugita: 70. Some sources say 80. Killed in action (KIA) 15 April 1945.
- Saburo Sakai: 64 (2 in China). Survived the war; died September 2000.
- Takeo Okumura: 54 (4 in China). Missing in action (MIA) 22 September 1943.
- Hiroyoshi Nishizawa: 36 official (102 claimed). KIA 26 October 1944.
- Toshio Ohta: 34. KIA 21 October 1942.
- Kazuo Sugino: 32. Survived the war; served in the Japanese Maritime Self-Defence Forces post-war.
- Junichi Sasai: 27. KIA 26 August 1942.
- Shizuo Isii: 29. KIA 24 October 1943.
- Kaneyoshi Muto: 28. KIA 24 July 1945.

Above: Japanese ace Junichi Sasai poses with an abandoned USAAF P-40 in the Netherlands East Indies, December 1941. (Source unknown)

Zero Survivors

Like many Japanese aircraft of the Pacific war, most surviving Zeros are made up of parts salvaged from multiple airframes. In many cases, such as those recovered after decades in a wrecked condition, they have been reconstructed to the point that the majority of their structure is made

An A6M3 Zero flown by Hiroyoshi Nishizawa, Japanese ace of the Pacific war and commander of the Tainan Wing.

up of modern parts. All of this means the identities of survivors can be difficult to confirm.

Most airworthy Zeros have had their engines replaced with similar American units; only one, an A6M5 held by the Planes of Fame Museum in Chino, California, is the only airworthy example with an original Sakai radial engine.

Airworthy Zeros are rare, one reason being that the light alloy used in the fighter's construction was susceptible to corrosion.

Right: Mitsubishi A6M5 Model 52s abandoned by the Japanese at the end of the war, captured by American forces at Atsugi air base along with many other types. (US Navy)

Captured Zeros being tested by the Allied Technical Air Intelligence Unit over Malaya, 1945. The South-East Asia TAIU was jointly operated by the RAF and USAAF. (IWM)

Photograph of a restored Zero taken from the window of a DC-3 on a flight from the Reno Air Races to Wisconsin. The mountain in the background is Peavine Peak outside Reno, and near the Stead airfield where the Reno Air Races are held. (Mark Grossman)

A6M2 Model 21 on display at the Pacific Aviation Museum, Hawaii. This aircraft was made airworthy in the early 1980s before it was grounded in 2002.

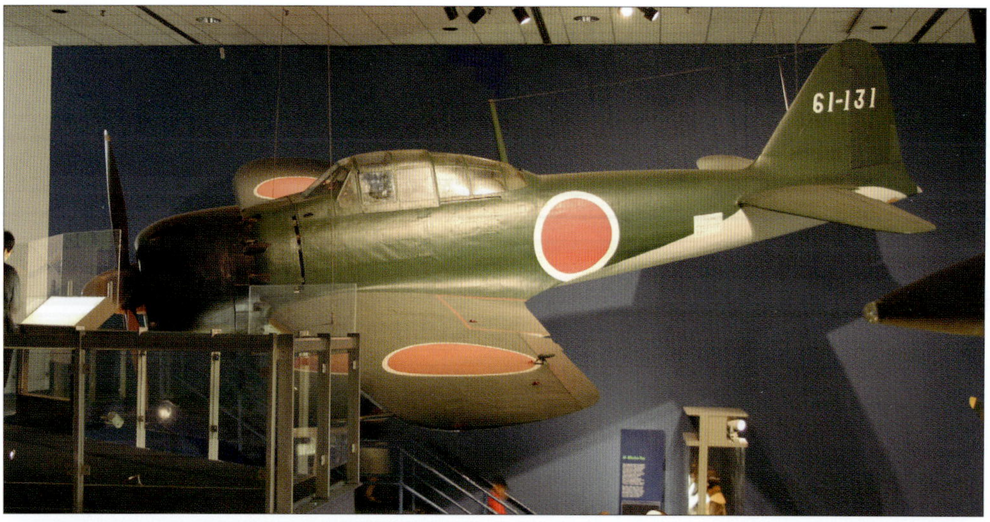

An A6M5 Zero on display at the National Air and Space Museum.

This Nakajima-built A6M2 was placed on display in 2004. It was found in Papua New Guinea, near the city of Kavieng on New Ireland, and was probably one of the aircraft delivered to Rabaul and operated at Kavieng by the 6th Kokutai (Squadron) and later by the 253rd Kokutai. It is painted to represent a section leader's aircraft from the aircraft carrier *Zuiho* during the Battle of the Bismarck Sea, March 1943, in which Allied air power won a major victory over Japanese sea power. American and Australian aircraft sank four Japanese destroyers, 15 troop ships and shot down more than 50 Zeros, preventing Japanese reinforcements from reaching New Guinea.

This A6M2 Zero, showing the engine installation, is an exhibit at the Museum of Nature and Science, Tokyo.

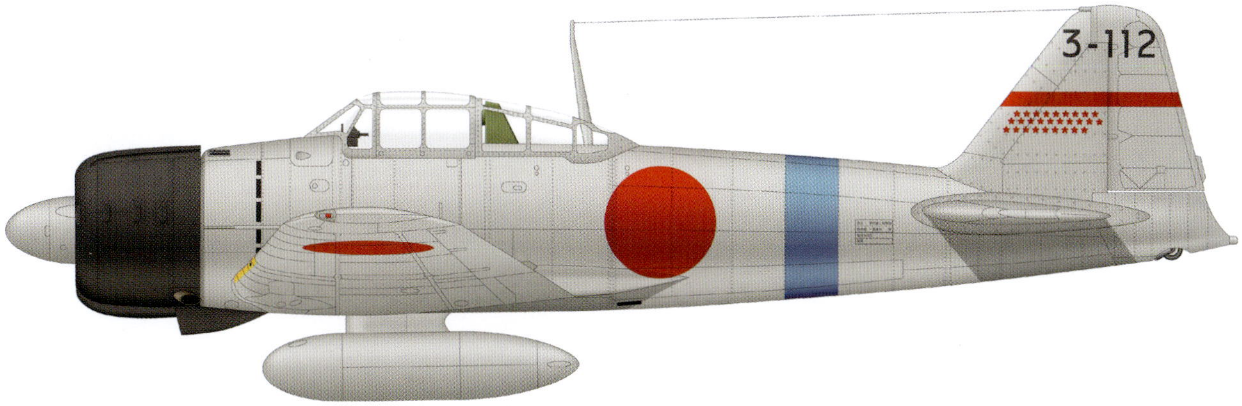

A6M2 Type 0 Model 11 flown by Lieutenant Minoru Suzuki of the 12th Kokutai, China, summer 1941. The 12th Kokutai (Air Group) was one of two established before the onset of the Pacific war, the other being the 11th.

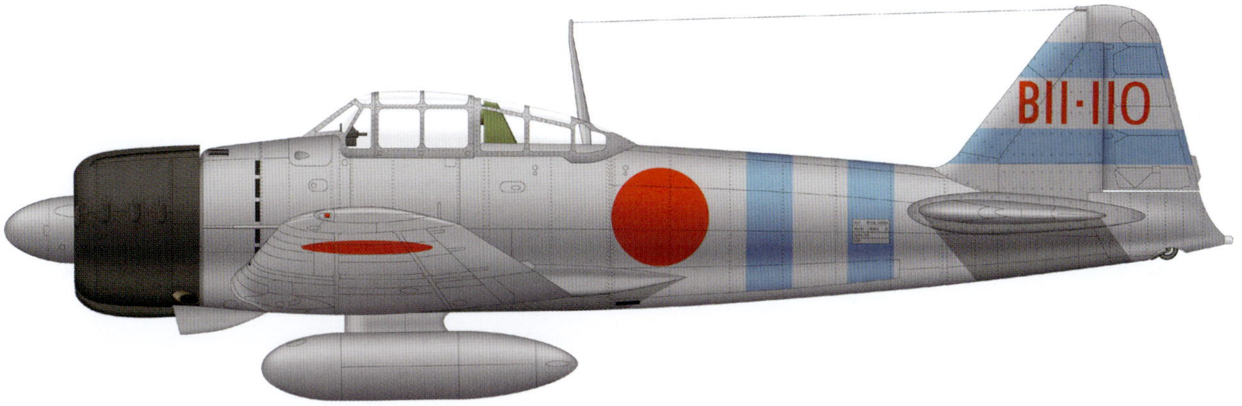

A6M2 Type 0 Model 21 of the 1st Koku Kantai, 2nd Koku Sentai, aircraft carrier *Hiryu* air wing, Pearl Harbor attack, 7 December 1941. In all, the Zero squadrons lost 15 aircraft out of a total of 79 taking part in the Hawaiian Operation.

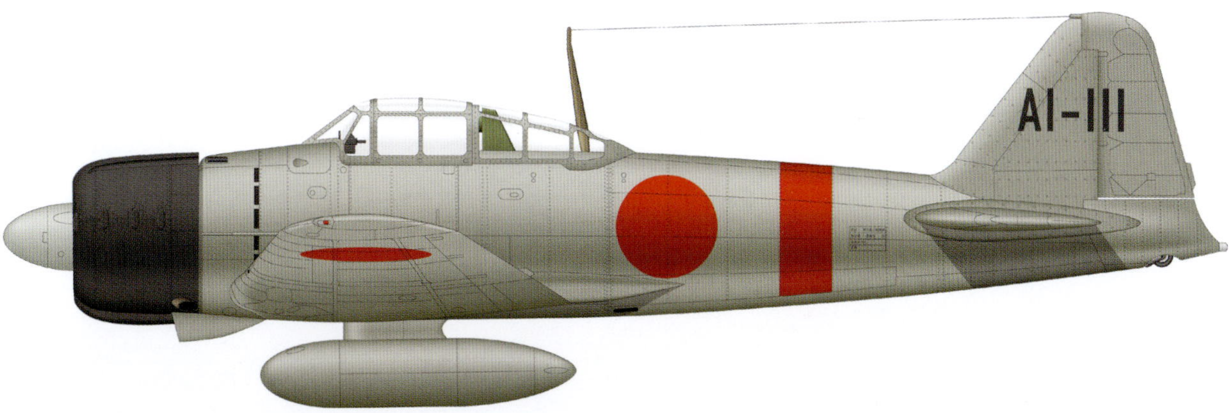

A6M2 Model 21 Zero, *Akagi* Fighter Squadron. Zeros of the *Akagi* Fighter Squadron were heavily involved in the attack on Colombo on 5 April 1942, claiming nine British fighters destroyed and seven probably destroyed.

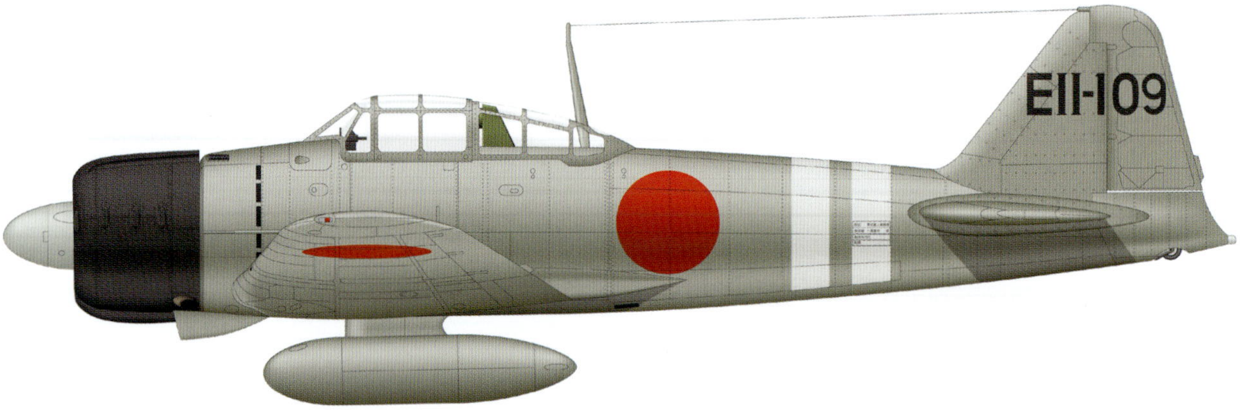

A6M2 Zero Model 21 of the *Zuikaku* Fighter Squadron. The unit insignia EII was used during the period December 1941 though October 1942.

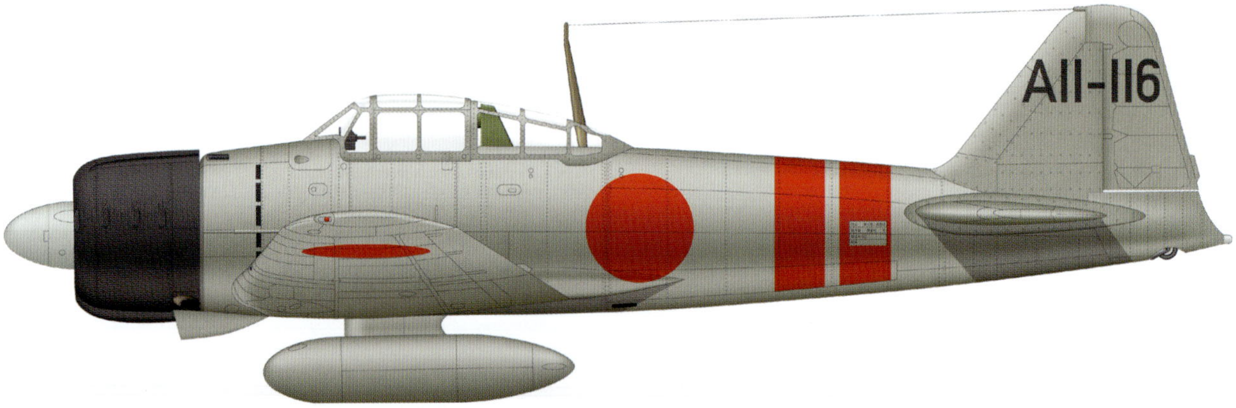

The two red bands around the rear fuselage of this AS6M2 Zero denote that it belonged to the aircraft carrier *Kaga*, of the 1st Carrier Division.

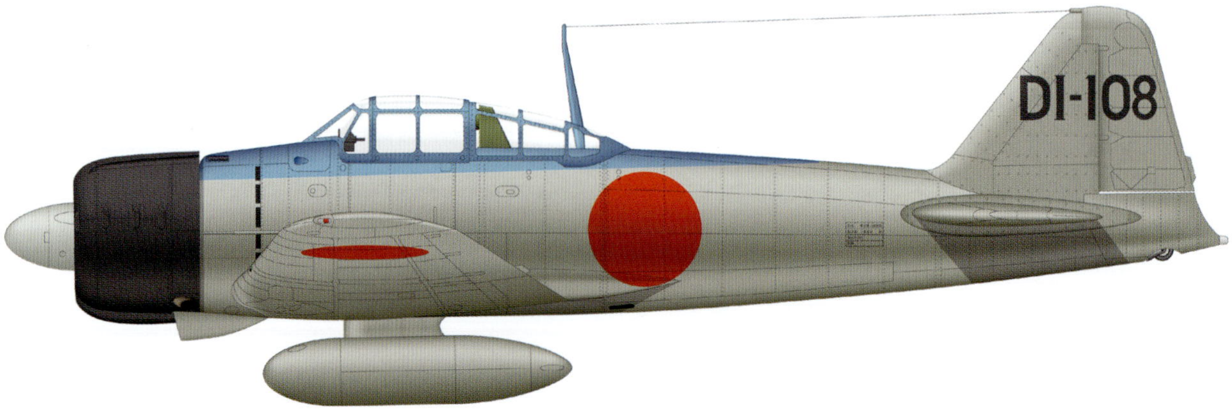

A6M2 Zero Model 21 of the *Ryujo* Fighter Squadron, flown by Petty Officer 1st Class Tadayoshi Koga during the June 1942 Aleutian Islands operation. The unit insignia DI was used between April 1941 and June 1942.

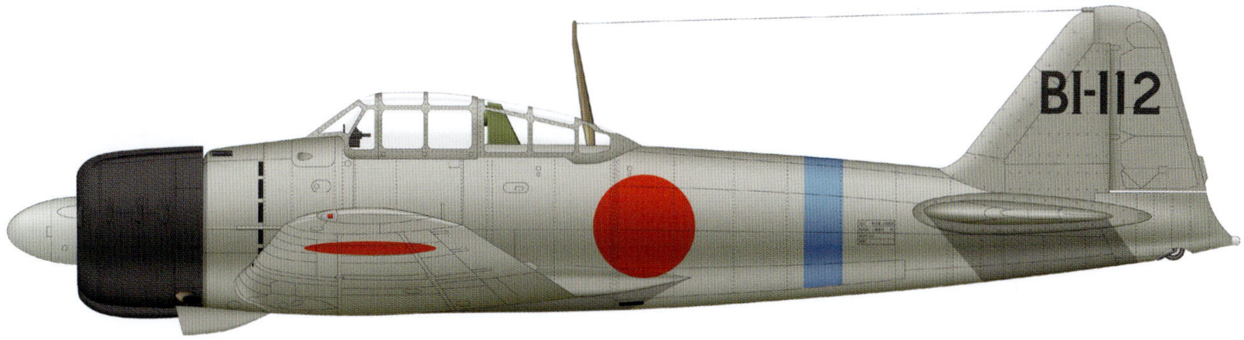

Mitsubishi A6M2 Zero Model 21 of the *Soryu* Carrier Air Group, engaged in the attack on Port Darwin, Australia, February 1942.

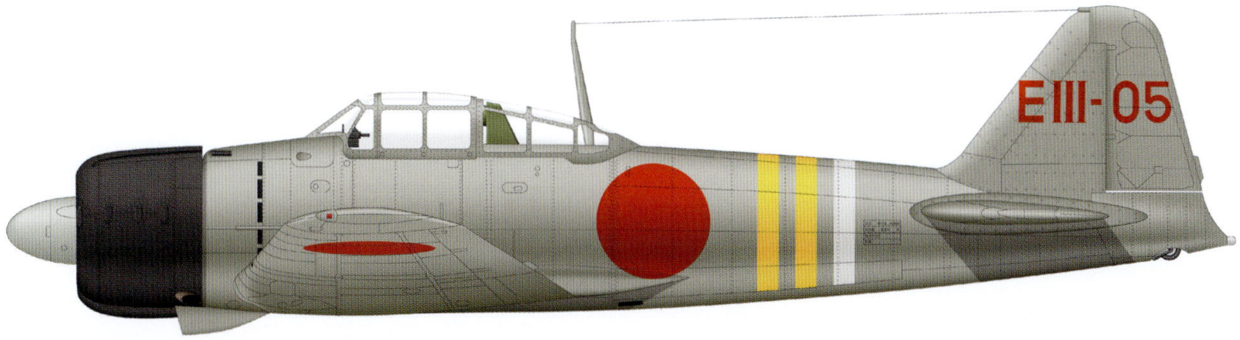

A6M2 Zero Model 21 of the *Zuiho* Fighter Squadron, as seen at the time of the Battle of Santa Cruz in October 1942.

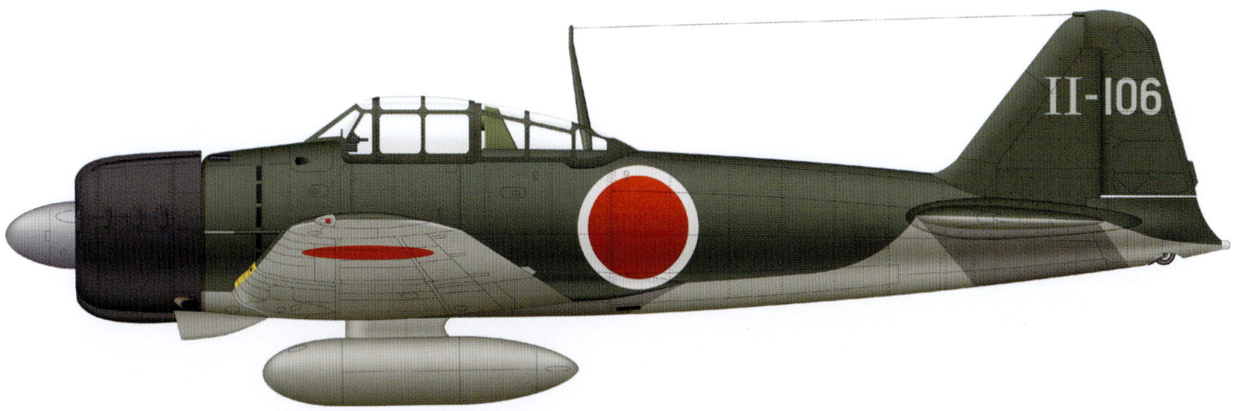

In July 1943, with most of the Zero units operating from land bases following the loss of the principal carriers, the IJN ordered that a scheme of dark green over light grey should be adopted, as seen here.

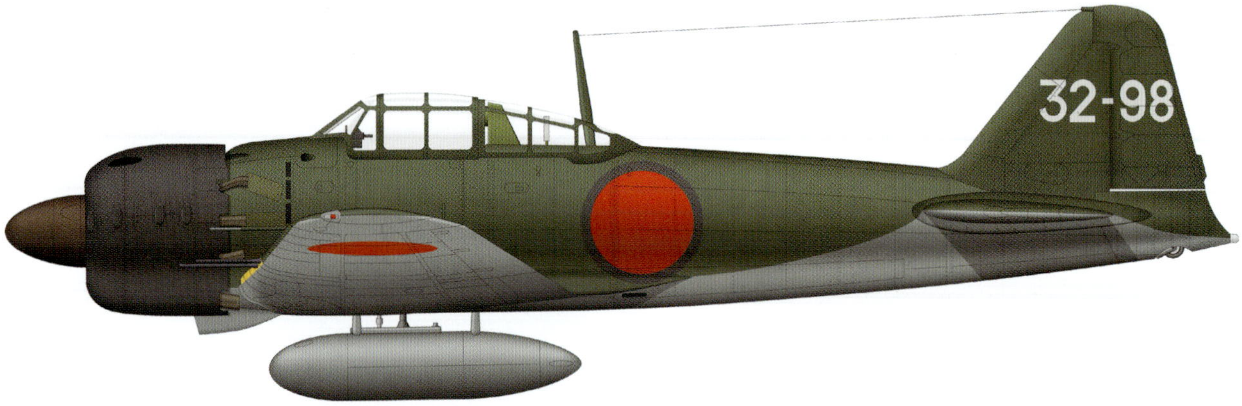

Carrier-based A6Ms of the early war period used an alpha-numeric system that identified their assigned carrier division, so the code AI in this system referred to the 1st Carrier Division's third carrier. Flight leaders were distinguished by bands on the fuselage or tail. This Rabaul-based A6M5 Zero carries a numeric tail code.

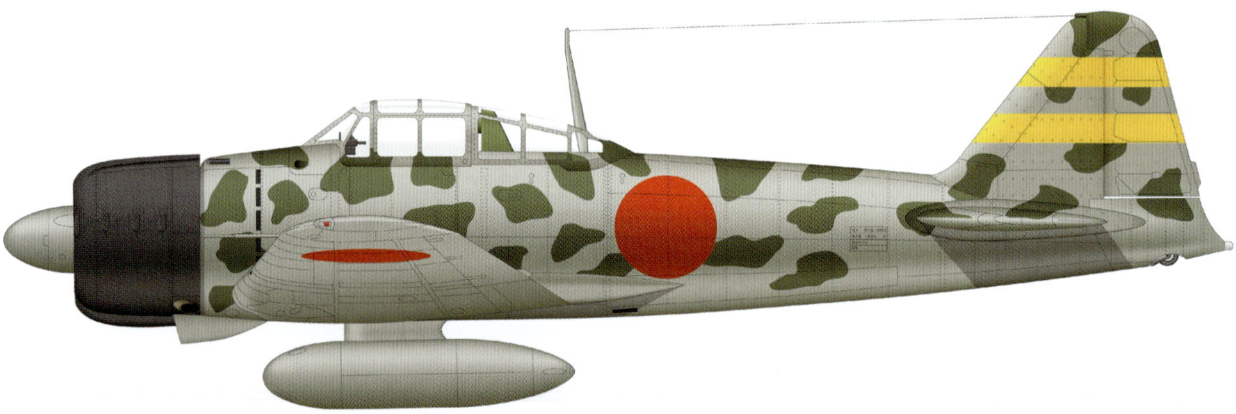

Although the standard camouflage scheme for land-based Zero units was dark green over light grey, individual commanders experimented with different schemes as their airfields were subjected to more intensive Allied air attacks. This A6M2 sports a dappled grey-green scheme. The two yellow tail bands indicate that the aircraft belongs to the *Shoho* Air Group.

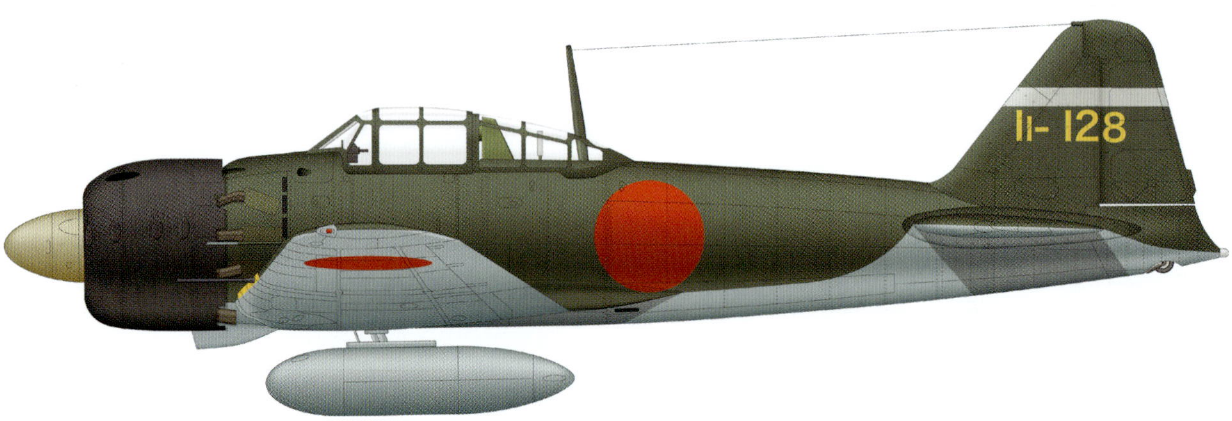

A6M3 Zero Model 22 of the *Zuikaku* Air Group as seen at Bougainville early in 1943. The original light grey colour scheme has been heavily smeared over with dark green.

Another example of an innovative camouflage scheme is displayed by this Zero, unit unknown. The red Hinomaru national markings were retained throughout the war, despite the fact that they stood out vividly against the camouflage background.

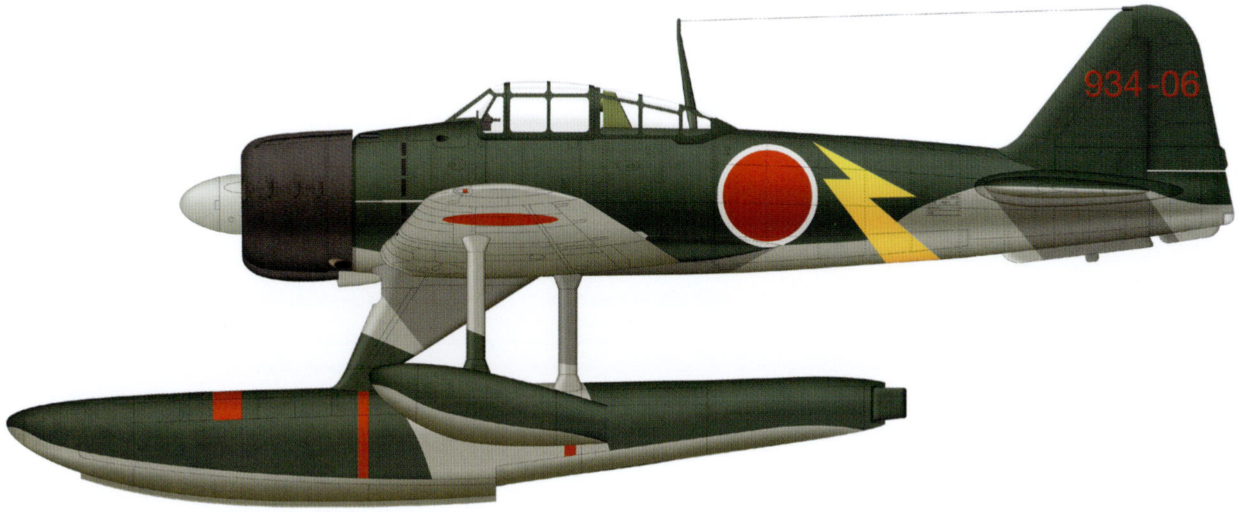

A6M2-N fighter floatplane, 934th Kokutai, Ambon Island, September 1943. The A6M2-N was first encountered during the battle for the Solomons. Despite its large float, creating extra drag, the aircraft was fast and manoeuvrable.

A6M3 Zero Model 22a of 582 Kokutai at Kahili airfield, 1943. Kahili Airfield, also known as Buin Airfield, was an airfield located near Buin, Bougainville Island, Papua New Guinea. It was neutralized by Allied bombing.

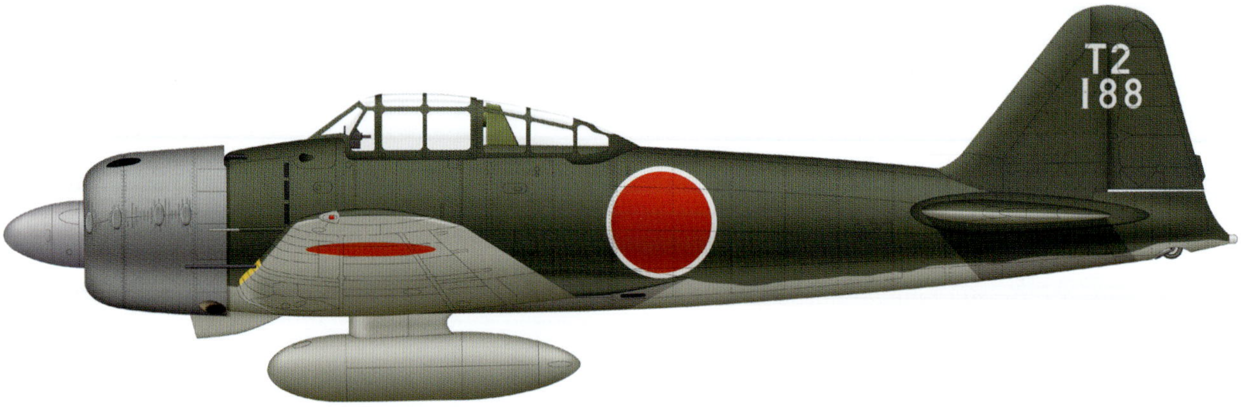

A6M3 Zero Model 32 of the 204th Naval Air Group, 1943.

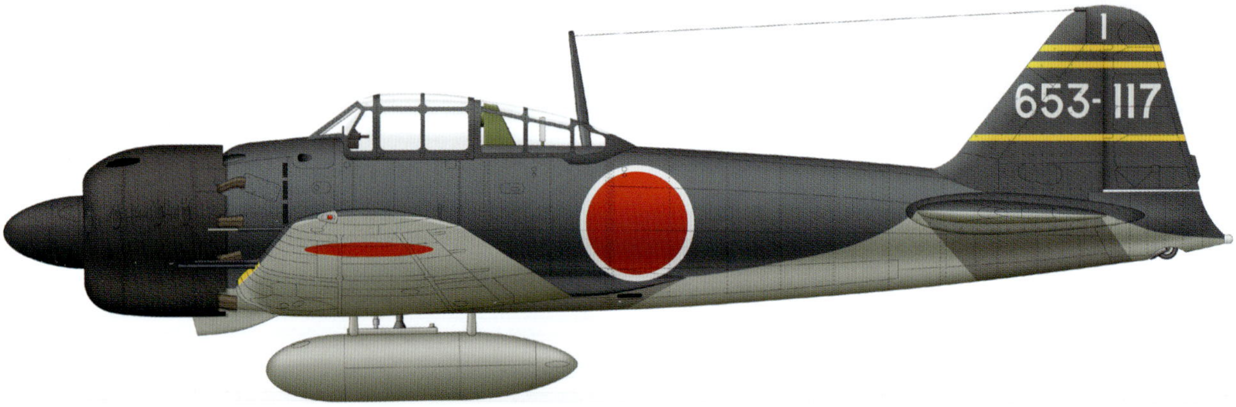

This A6M5 Zero of the 653rd Kokutai saw action in the Battle of the Philippine Sea in 1944, operating from bases on land.

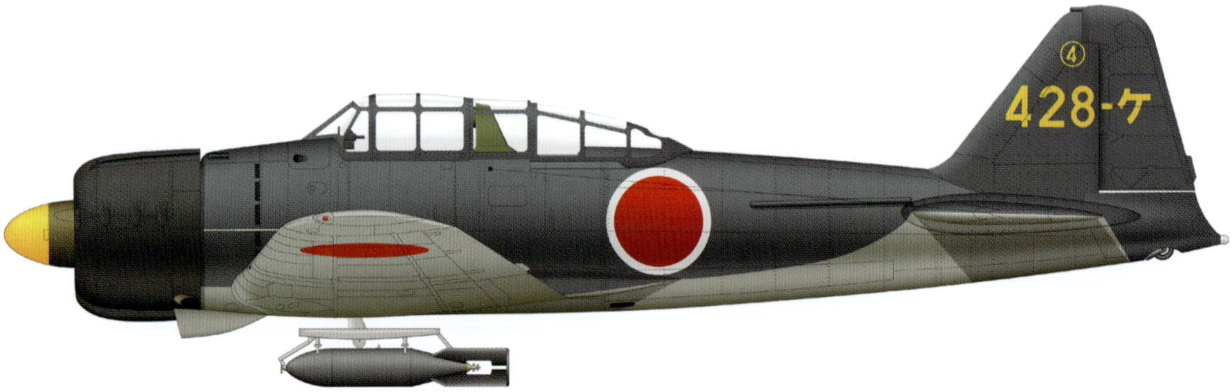

Operational units improvised a bomb rack for the Zero that allowed it to carry a 250kg bomb in place of the drop tank. This fighter-bomber concept was taken up by the manufacturer as the A6M7, which had a centre bomb rack and fittings for two drop tanks outboard of the wing armament.

A6M2-K Zero trainer of the Tsukuba Training Kokutai as seen in 1944. Japanese training aircraft were painted bright orange overall except for the engine cowling, which remained black.

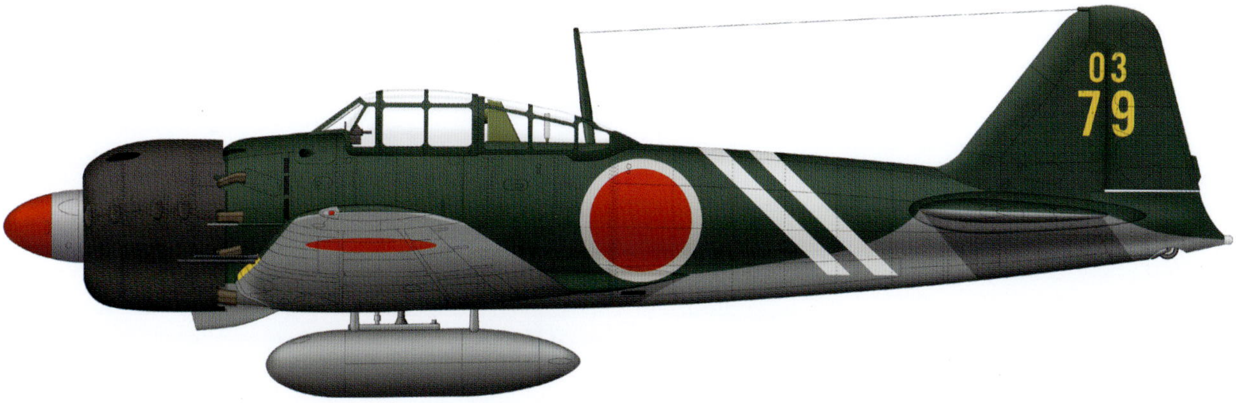

A6M5c Zero Model 52, of Hikotai 303, Kokutai 203, Kagoshima Air Base, Japan June 1945. This aircraft was flown by Warrant Officer Taskeo Tanimizu, an ace with 18 victories.

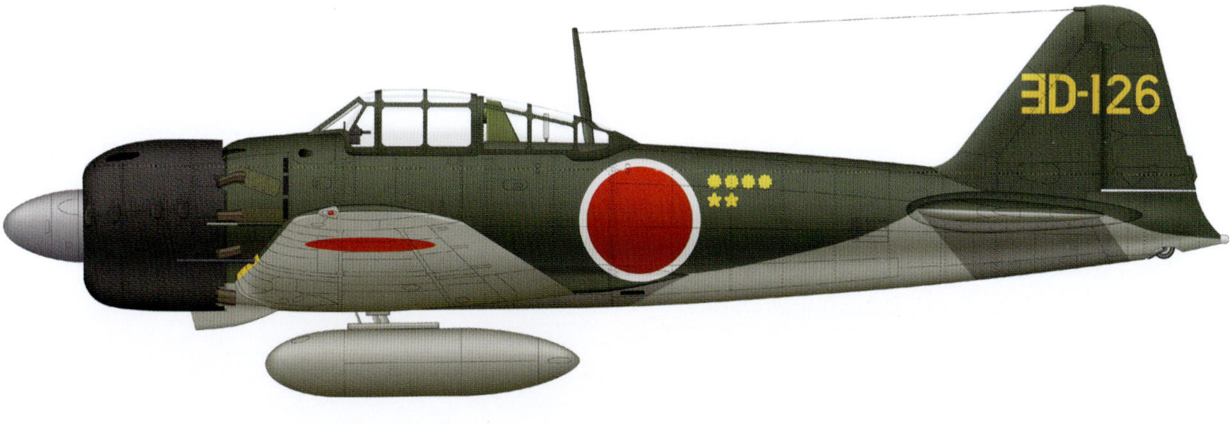

This A6M5 Zero was flown by Ensign Sadaaki Akamatsu of the 302nd Air Group, who scored six victories.

Mitsubishi Zero 341-H-81 of the 341st Kokutai, 401st Sentai, Manila, 1945.

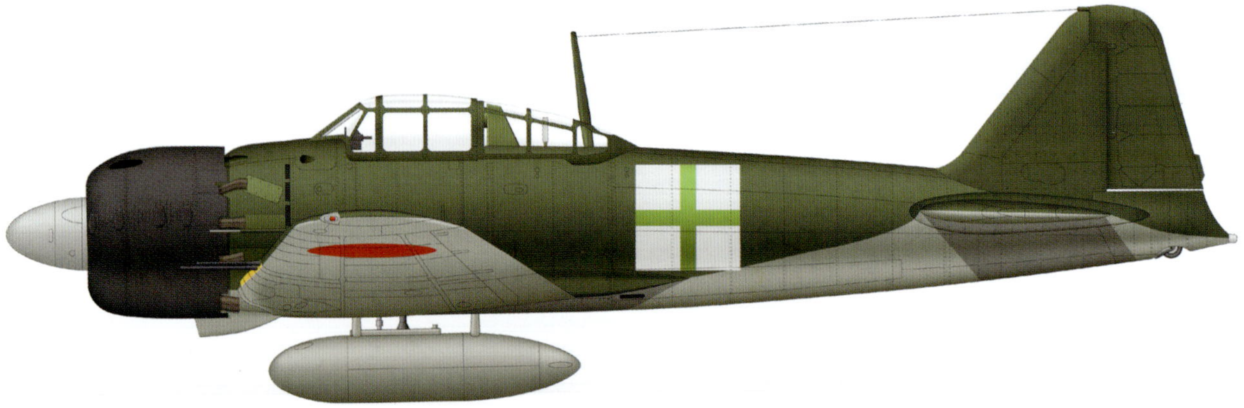

A6M5c Model 52 Zero in surrender markings, 1945.

A6M2 Model 21 Zero in Allied Tactical Air Intelligence unit markings.

Modelling the Mitsubishi Zero

The early Airfix kit dates from 1959 and went through all of the various packaging types from bags to boxes.

The box art was the best feature of the 1963 Revell kit, although it was marginally better than the Airfix.

Such is the historical significance of the Zero, it is hardly surprising that it has been represented in kit form many times in all the major scales – as well as some of the minor ones. About 70 different representations are known to me, with a large number of re-issues/re-boxings included in that number. Clearly it would be both impractical and rather tedious to examine all of these. Many are very rare and often the quality and accuracy of the earlier kits leaves so much to be desired that they can only be considered curios today. However, many silver-haired modellers will have fond memories of some of these early kits, so I've taken a look at the ones that are most commonly seen on the UK second-hand market these days.

The **Airfix** A6M2 Zero was first issued in 1959. It remained in production until well into the 1990s, and it took until 2011 for a newly tooled version to be issued, which will be examined later. The Zero appeared in the full range of Airfix packaging over the years, from the early bags through to a special-edition box in 1990. Even when it first appeared it was an unimpressive kit. In terms of accuracy, length and span are about right and the wing and tailplane shape in plan view are reasonable. Sadly, that's the end of the good news. The canopy is grossly overscale (nearer 1/64) and the fuselage is far too wide in order to accommodate this. The undercarriage doors and wheel wells are incorrectly shaped. All of the smaller parts are crude and overscale and the surface detail consists of boilermaker-sized rivets and entrenched panel lines. The cowling is a rough moulding of approximate accuracy. The fit of parts is generally poor, with wing–fuselage, tailplane–fuselage and cowling–fuselage being particularly bad. At least it does well in the nostalgia ratings.

The early 1960s saw the appearance of the **Revell** A6M5 which, like the Airfix offering, remained in production for a long time and went through a variety of re-boxings. In terms of accuracy, the basic dimensions are good, but the wing plan is marred by the chord being too narrow at the root, which means the taper from the tip to root is not sharp enough. Tailplane plan is good. The fuselage profile is reasonable but the fin shape is not quite right. The canopy is fairly accurate and well-formed, as is the droptank. The worst feature of the kit is the awful cowling, a rough moulding that bears no relation to anything that was ever fitted to the front of a Zero. The error is compounded by a poor propeller, which has blades moulded as if it rotates clockwise, and inaccurate exhaust stubs. These are of the incorrect pattern and number and are moulded integrally with the fuselage halves. The kit is covered with heavy rivet detail and some deeply entrenched panel lines. Fit of parts is not exceptional, but better than the Airfix kit. The box art on the first issue is great though.

The last member of the old guard to be examined is the **Frog** A6M5 from 1961. This also appeared in various re-boxings, although it didn't remain in production for as long as the Airfix and Revell offerings. Again, accuracy is a bit of a disaster, with the wingspan 3mm over and a tailplane that is too narrow and pointed. The fuselage profile is quite good but the fin is undersized. The overall appearance is

The Frog kit was another 1960s offering that's high on nostalgia but low on accuracy.

The Matchbox kit from 1973 was a clean moulding but offered little improvement over its predecessors.

spoilt by a hopelessly inaccurate cowling and a 'squashed' canopy with a very sharply raked windscreen. The cowling has a curiously humped top when viewed head-on. The air scoop is totally wrong as is the shape of the gun ports. None of the distinctive exhaust stubs is included at all. The undercarriage legs are thick, heavy mouldings and the wheel bays are moulded solid. Surface detail consists of heavy raised lines of incorrect pattern. Fit of parts is better than the Airfix kit, but that's about it for the positives.

Despite the limitation of these kits, it took a long time for anything that could be considered a significant improvement to appear. **Matchbox** entered the plastic kit market in the early 1970s and produced a Zero in 1973. It was more cleanly moulded, with better fitting parts and superior decals to the products previously described, but in terms of accuracy it was actually worse. It was marketed as an A6M2 but was something of a hybrid between an M2 and an M5, with a completely inaccurate cowling thrown in. Partial redemption came from Japan. There had been Japanese kits of the Zero available for some time, but the 1960s was very much a fledgling time for the plastic kit industry in that country, which had yet to achieve the dominance with which we are so familiar now. Zero kits from the likes of Nichimo and LS were sometimes seen in the UK, but in terms of accuracy they were every bit as poor as the contemporary European offerings. Things began to change rapidly in the 1970s though.

In about 1973, **Hasegawa** released separate kits of the M2, M3 and M5. The M5 version has overall dimensions that are spot on. Wing and tailplane shape in plan view are very good, although the elevators are not accurately scribed. The ailerons are far too narrow on the wing underside, with heavy rib detail and an inaccurate trim tab. The propeller, spinner, bombs, drop tank and undercarriage legs are well represented. Wheels and engine are reasonable. The problem areas are the cowling and canopy. The cowling is too straight-sided, with none of the subtle curvature of the real thing. The cowling gills are poorly marked and the exhaust stacks not very convincing. The canopy scales out well in plan and elevation, yet it has a rather pinched appearance which doesn't quite look right. Surface detailing is a bit coarse, especially on the wings, and the fit of parts is mixed, with the wing–fuselage joint being particularly poor. Despite the faults, these kits were a significant step forward over their predecessors. We still awaited an offering that could be considered definitive though.

We had to wait until the 21st century. Having said that, the Korean firm **Academy** did produce an A6M5 in 1998. This is still around and it did offer some improvement over the aforementioned kits. In particular, improved overall accuracy and better cockpit detail. A rather thick, clunky canopy and unconvincing exhaust stacks were against it though. And of course we now – at last – have some genuine quality available. Airfix, Tamiya and Hasegawa all currently produce kits on the Zero that are a vast improvement over anything in the scale that existed before. All are the result of completely new tooling. So how do they compare?

It took **Airfix** a little matter of 52 years to get round to improving their 1959 moulding. The new kit is the A6 M2(b) and appeared, I believe, in 2011. It has already been reissued in at least three different boxings since then but the kit is unchanged. In all respects, it is on a different planet from the original kit, and studying the two side by side provides a perfect example of how the kit industry has developed during that time. The latest kit is very cleanly moulded with a good level of detail for its scale. There is an excellent decal sheet. The cockpit is well done (better than Hasegawa although not as good as the Tamiya kits) and so is the engine. Wheel bay detail is also quite effective. Again, it scores over the Hasegawa in this respect. General outline accuracy is virtually faultless as far as I can tell, with the possible exception of the canopy. This appears to be a bit too wide – it's certainly wider than either the Tamiya or Hasegawa offerings but this is barely noticeable on the finished model. Another issue with the canopy is the frame lining which looks a bit too broad, particularly on the rear section. The most annoying feature with this new kit is the overdone surface panel detailing. The panel lines are very crisply moulded but look too deep. This is something of a matter of personal taste, and I know that a lot of modellers today go to lengths to emphasise panel detailing on finished models and would be quite happy with this aspect of the kit, but to me it's not accurate and the Japanese kits do it better. Grumbles apart, this is a good kit that can be made into an attractive model. It's also cheaper than its Japanese rivals.

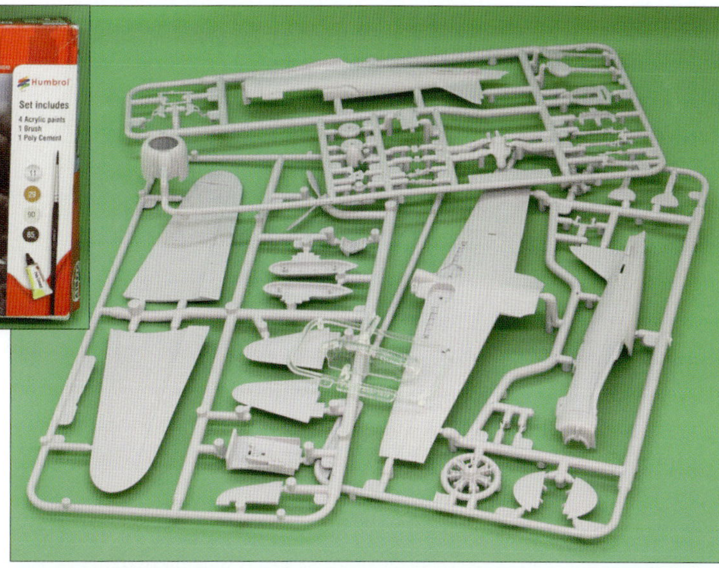

Above: The 2011 Airfix re-tool is a most presentable kit.

Right: The spread of parts for the Airfix kit. Mould quality is very good.

Below: The new Airfix kit is let down by the overdone surface detailing, seen here on the lower wing.

It builds into a nice model. These examples by Peter Terry of the Japanese Aircraft SIG are finished in after-market decals.

Two examples of the new Hasegawa 1/72 kits. They produce all the main Zero variants.

The newest **Hasegawa** Zeros predate the latest Airfix kit by some years. Hasegawa have released all the major variants in different boxes but based on a common set of moulds. I looked at the A6M2 and the M5(c). The outline shape and general accuracy is hard to fault. In fact, the major components of the two different brands are virtually interchangeable. The advantages offered over the Airfix kit by the Hasegawa one are a better canopy, which is a slightly finer moulding with finer (but still a bit broad) frame lines, and a more extensive decal sheet. Most important to me, it has better surface detailing with the finely engraved panel lines looking much better than the overdone Airfix rendering. The kits assemble well, with just a bit of attention required around the insert that fits on top of the fuselage in front of the canopy. Price wise, it's dearer than the Airfix but cheaper than the Tamiya.

That leads us on to the **Tamiya** kits. These are the most recently produced (the date on my sample is 2013) and the most expensive. As with Hasegawa, Tamiya produce all the main A6M variants in different boxes, with a lot of parts in common. Compared to the previous two offerings, I have to say that the Tamiya has all their good points but none of their drawbacks. It has even better cockpit detail than the Airfix and just as good a

Hasegawa A6M2.
(Frederick Seitz)

Hasegawa A6M2.
(Frederick Seitz)

BII-187

representation of the wheel bays. It has the better canopy of the Hasegawa, in fact even more so as it offers a multi-piece option to allow an open hood to be displayed. It also has the best representation of the canopy frame lines. It has beautifully refined engraved surface detailing and is the sharpest, crispest moulding of them all. Assembly is straightforward and the decal sheet is also very good.

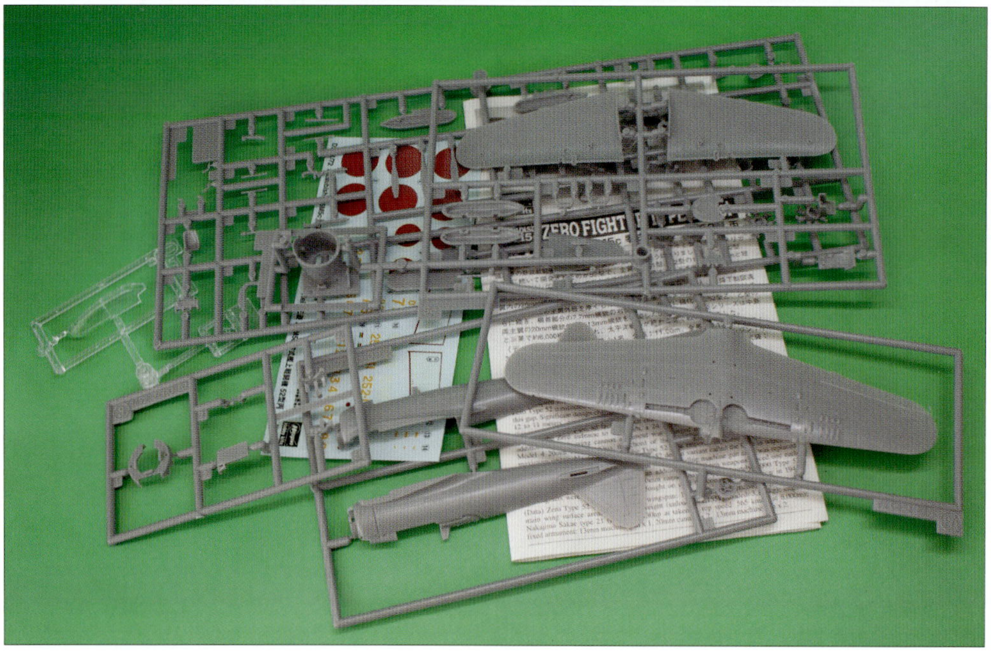

The spread of parts from the Hasegawa 1/72 A6M5. It's a very tidy kit that's hard to fault.

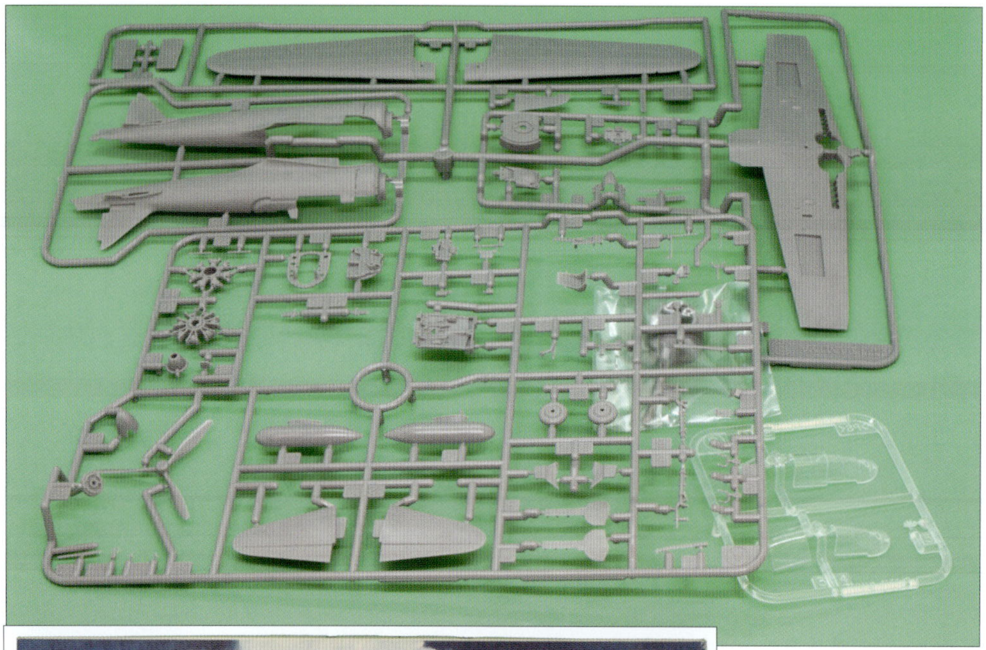

The new Tamiya kit is the best of the lot, but it's also the most expensive. As with Hasegawa, they produce all the main Zero variants.

Above: The Tamiya 1/72 A6M3, another model by Peter Terry. The Zero is a great subject for those modellers who like their weathering!

Left and below left: Tamiya publicity photos of the 1/72 M2 and M5 variants. (Images courtesy of Tamiya Inc.)

Whilst it took a long time for the plastic kit industry to produce a worthwhile Zero in 1/72 scale, it was a different story when moving up to 1/48. The earliest item in this scale was the **Monogram** A6M5 of 1962. In its day, this was probably the best Zero kit available in any scale. It was re-issued several times and samples are still common. In terms of accuracy, basic dimensions were good but the ailerons were incorrectly represented, as were the flaps. The tailplane plan was quite good although the tips were a bit too pointed. The fuselage was generally good but the fin was too small and inaccurately shaped. The canopy was a bit thick but of the correct size and shape. The exhaust stubs were a separate moulding and looked quite effective, but the bottom exhaust outlet (a twin pipe) on each side had been omitted. The propeller and spinner were well moulded and accurate, but the engine was moulded integrally with the cowling and looked unconvincing. The main weakness of the kit was the cowling. Despite having well-formed gun ports, its overall shape was suspect. It appeared too straight-sided and lacked the full curves of the real thing. The undercarriage was designed to retract and lost accuracy as a result. Also, no cockpit detail was provided, other than an inaccurate instrument panel. The kit was a clean moulding and assembled well. Obviously it has been outclassed by much later offerings but was a worthy effort from probably the best kit manufacturer of the time.

A 1960s kit built in the 1960s way! The Monogram 1/48 A6M5 built by Jonathan Prestige. He keeps it as a reminder of 'progress'.

3D-150

3D-150

3D-150

3D-150

3D-150

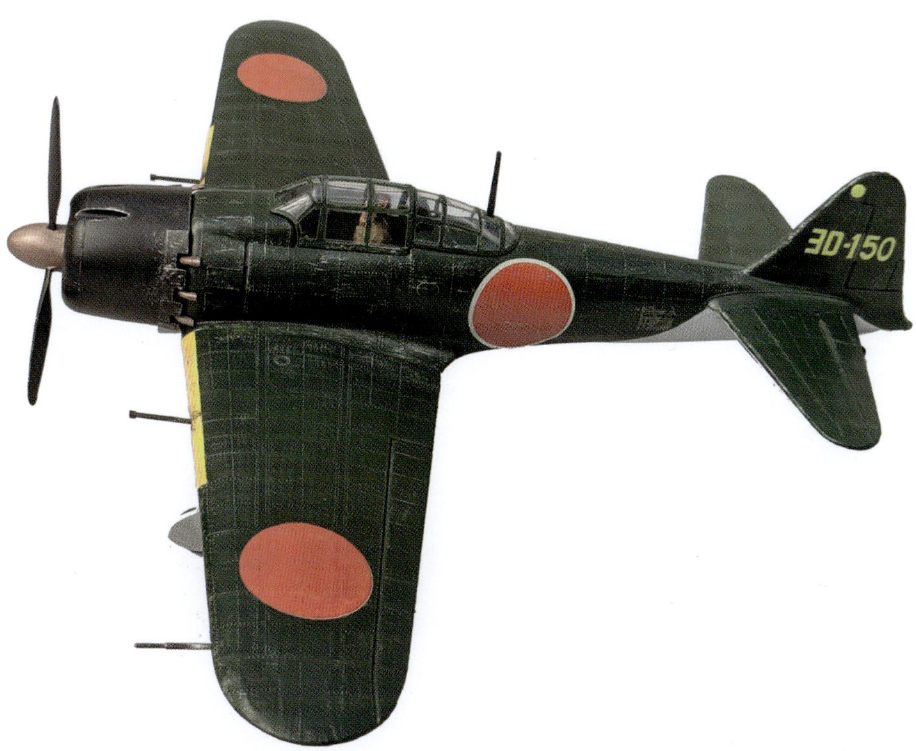
3D-150

The very worthy Otki/Arii kit from the early 1970s.

By the 1970s, the Japanese kit industry had really found its feet and this is no more evident than in the 1/48 Zero kits they produced at that time. Two of note are the **Nichimo** A6M5 and the slightly later (1974) **Otaki** offering of the same variant. The Nichimo kit addressed some of the shortcomings of the Monogram effort but was seldom seen in the UK, unlike the Otaki (later released under the Arii label) which is still around and can often be obtained at knock-down prices. The Otaki kit has always been a favourite of mine. It represented a quiet but significant development in Japanese kit manufacturing. Gone were the working features and motorization provision that had blighted the accuracy of many of the previous kits. Instead, we have a fairly simple model where the focus seems to be on accuracy rather than toy-appeal. Although slightly short on overall dimensions (by 4mm on the wings and 2mm in length) and with wingtips that are a little too rounded, overall accuracy was outstanding for the time. A particularly good job was done with the awkward upper fuselage area in front of the cockpit. This was done even better than on the later and generally better Fujimi/ Tamiya/Hasegawa kits. The tricky cowling was represented well, and an attempt was made to provide detailing in the main wheel bays, although these were too shallow to look really convincing. A very good instrument panel was provided for the cockpit, but the rest of the detailing there was very basic. Surface detailing was engraved and included a lot of rivets but the kit was very cleanly moulded with parts that generally fitted well. Even today, one of these kits, built and finished to a good standard, captures the feel of the Zero very well.

The next group of 1/48 Zero kits appeared in the 1980s, and remained market leaders for the next 20 years. All are still available, either currently in production or readily available second hand. These represented a big improvement over any Zero kits in any scale that had gone before and still hold up well today.

First up, the **Fujimi** A6M2. Very good overall accuracy although the rear fuselage in profile looks a shade deep. The tricky section immediately in front of the cockpit does not quite have a sufficiently pronounced slope, but otherwise there is little else to fault. The separately packaged one-piece canopy is well moulded and the cockpit is very well detailed by the standards of the time. Undercarriage is well formed and accurate, with detailed wheel bays. Surface detailing is fairly restrained although some of the raised panels are a bit overdone. A notable feature of this kit was an excellent quality decal sheet that offered no fewer than eight marking options.

Next, to **Tamiya**, who produced all the major Zero variants, including the Rufe seaplane, in the early 1980s. Rather than look at them all, I shall focus on the A6M2 as it provides a direct comparison with the Fujimi kit just described. This Tamiya offering has almost spot-on accuracy, with the only problem area being the upper fuselage decking in front of the cockpit. This slopes too steeply, as opposed to the Fujimi where it doesn't slope enough. Notable pluses are the two-piece cowling which captures the deceptive curves of the original very well, very good cockpit and wheel bay detailing, and an excellent canopy. This comes as a one-piece moulding for a closed canopy option, but an alternative three-piece set can be used if an open canopy is preferred. The kit comes with aileron mass balances which were only fitted to earlier type 22s (up to No. 127) and were thereafter deleted. Only two of the aircraft covered by the decal sheet had these, and careful checking is required as the instructions are not that clear over this point. Probably the main fault with these Tamiya kits by today's standards is the raised surface detail. This is very restrained and generally nicely done, but is not to the taste of most current modellers. Compared to the Fujimi offering, this Tamiya one has the edge for me in terms of its overall moulding finesse, the optional open/closed canopy, and just a shade more detail.

In 1988 **Hasegawa** released the A6M5. Again, we find almost flawless accuracy, with the exception of the same problem in front of the cockpit – Hasegawa have sided with Fujimi over this, providing one with a slope that isn't quite steep enough. The cowling is a superb moulding, and the three-piece canopy is also very good. Engine, cockpit and undercarriage detail were excellent by the standards of the time, although not quite up to the standard of the Tamiya M5(c). At the time, the Hasegawa was the only Zero in any scale to come with the late-pattern wooden drop tank with fins, which was represented well. Surface detailing is a mixture of fine raised lines but also gentle engraving and is therefore possibly more acceptable to today's aficionados. A very good companion to the Tamiya A6M5(c). The kit has been reissued in various guises since its first appearance and, while no longer in production, is not difficult to obtain.

These kits were rightly considered the market leaders for a long time. As the after-market product industry developed, the kits were well provided for and could be upgraded in various ways to keep up with changing times and demands. Even now they offer a cheaper alternative to the newer kits, particularly suited to inexperienced modellers, or the more expert who don't mind doing a bit of extra work.

Time moves on, however, and both Hasegawa and Tamiya were to offer new toolings of their Zeros that are the ones we find on the shelves today. First Hasegawa, who issued their latest version in 1994. As with the 1/72 kits, the same basic mouldings were used to represent all the main Zero variants. Although going through various re-boxings and the additions of some new parts where appropriate, the basic kit is the same. Compared to the 1970s effort, the new one offers more refined detail, particularly in the engine and wheel bays, and to an extent with the cockpit. Surface detail has also been refined and the inaccuracy in front of the cockpit has finally been rectified by the insertion of a separate piece to capture the deceptive contours of that difficult area.

Interestingly enough, all the newer kits in all scales have gone for this solution, with equally good effect. The kit assembles easily and well. Possibly the most interesting of the Hasegawa releases came in 2008 when they issued an excellent representation of the A6M8 Type 54/64. This rather enigmatic variant was intended to improve the performance of the Zero by fitting a new engine, the Mitsubishi Kinsei Model 62. The result was an extensively modified nose and cowling which significantly changed the appearance of the aircraft. Only two units were completed, a prototype and a pre-production model, before the war ended. However, no Zero collection is complete without one. Unfortunately, this particular kit seems very hard to obtain at the moment in the UK, my sample coming directly from Japan. It seemed we had the definitive Zero kit at last. The situation changed significantly in 2008 however, when Tamiya released a completely new A6M5.

The box may be tatty when you come across older kits, but the contents of the first Hasegewa 1/48 Zero made an excellent kit for its time.

Hasegawa's new tool 1/48 A6M2 in captured Chinese markings. The Chinese fitted distinctive panels behind the cowling, evident on this model by Gary Wenco of the Japanese Aircraft SIG.

The Hasegawa A6M5 by Bill Anderson of the Japanese Aircraft SIG.

In contrast to his 1960s Monogram Zero, Jonathan Prestidge demonstrates his mastery of modern finishing techniques with this lovely example of the new Hasegawa A6M2b.

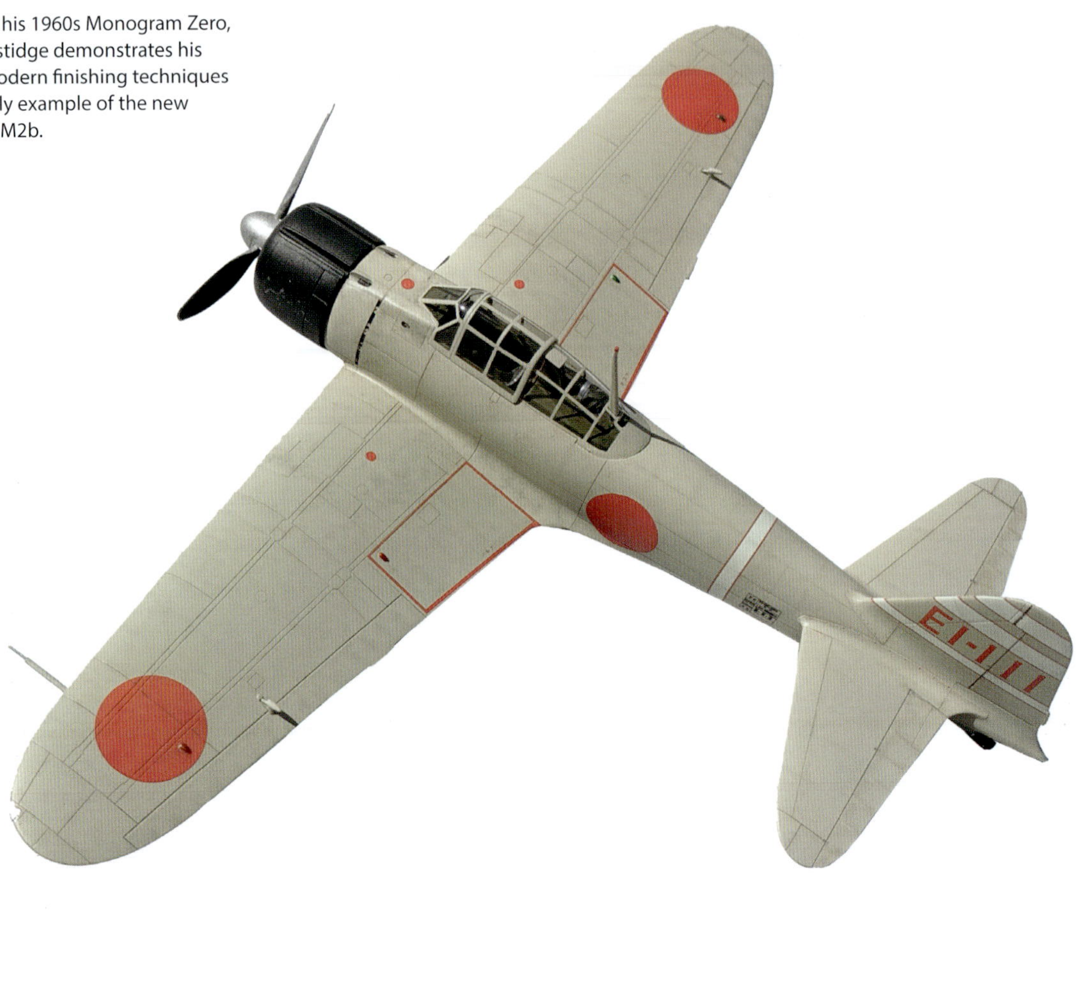

Hasegawa A6M2b.
(Images courtesy of Jonathan Prestidge)

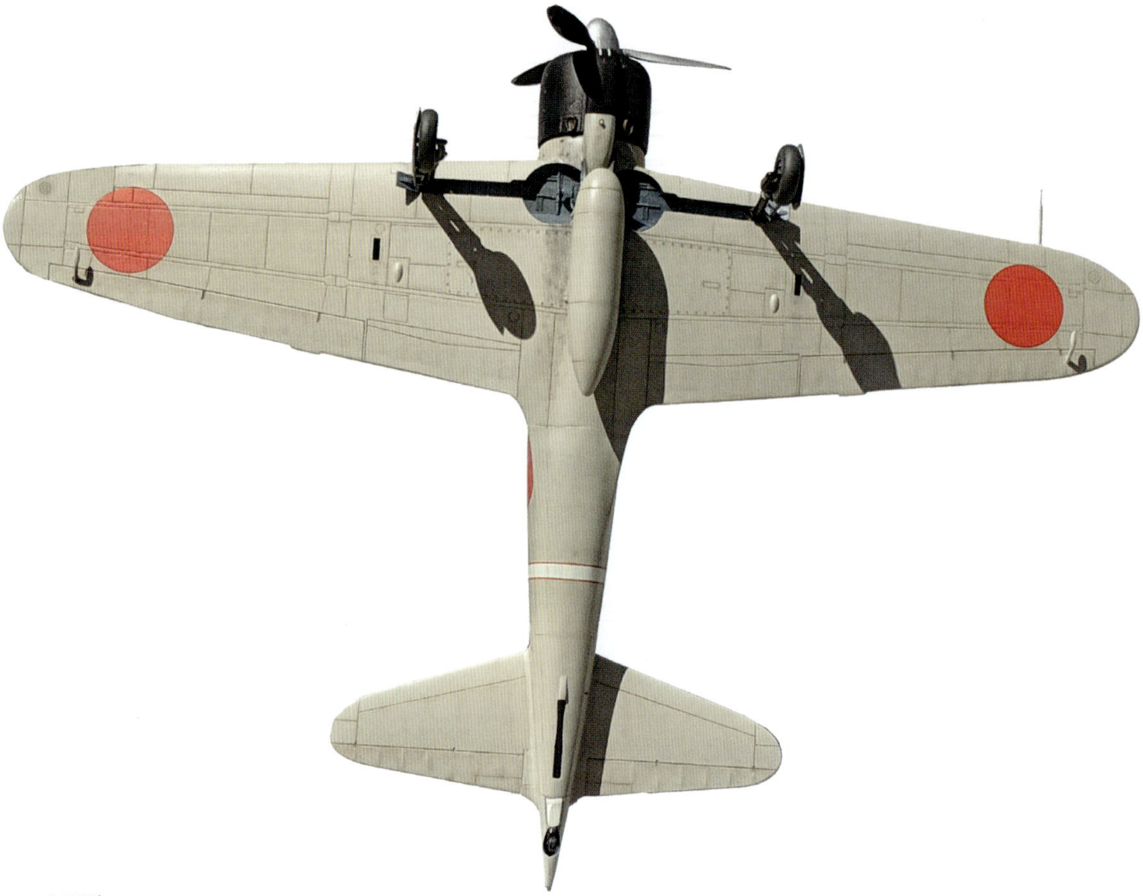

Hasegawa A6M2b.
(Images courtesy of Jonathan Prestidge)

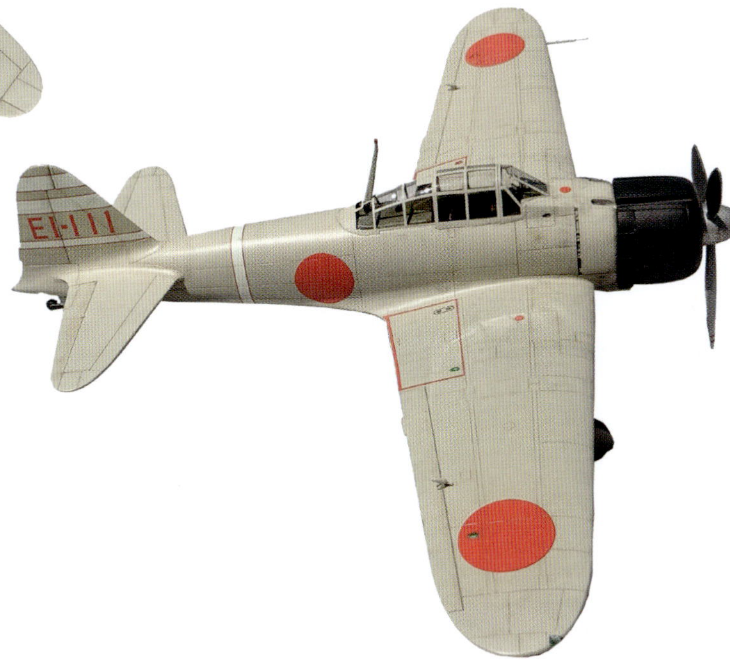

Hasegawa A6M2b. (Images courtesy of Jonathan Prestidge)

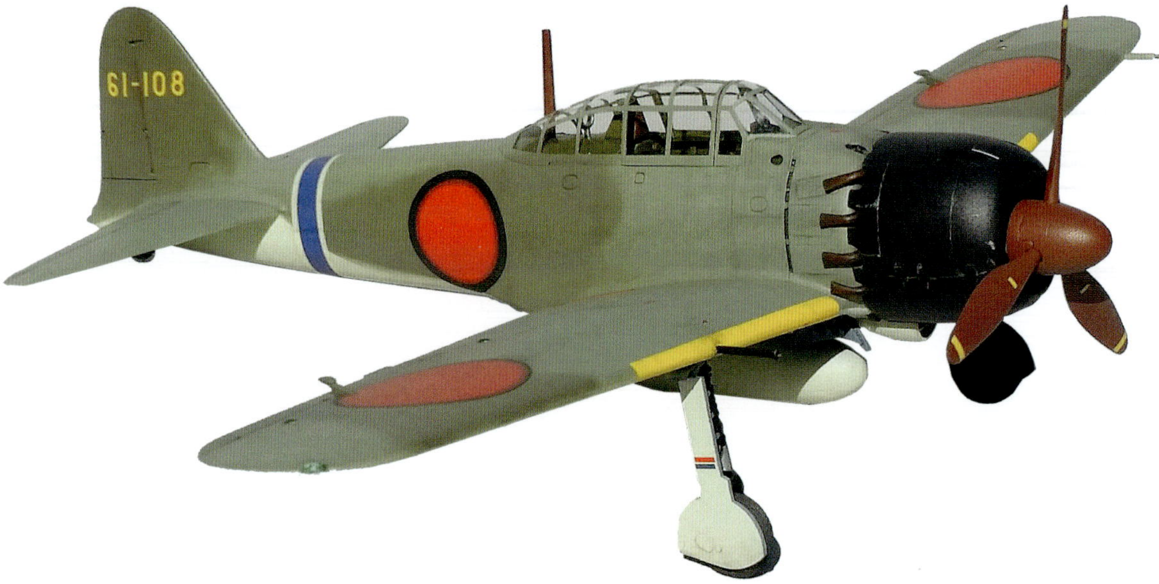

Another example of the Hasegawa A6M5, built by Jonathan Prestidge.

Another example of the Hasegawa A6M5, built by Jonathan Prestidge.

Another example of the Hasegawa A6M5, built by Jonathan Prestidge.

The last Hasegawa 1/48 Zero was the A6M8 in 2008.

The kit includes a large number of parts …

… but these are the most important: the new engine/cowling components. Hasegawa had an earlier attempt at representing this variant by including these parts in resin. This version is much better.

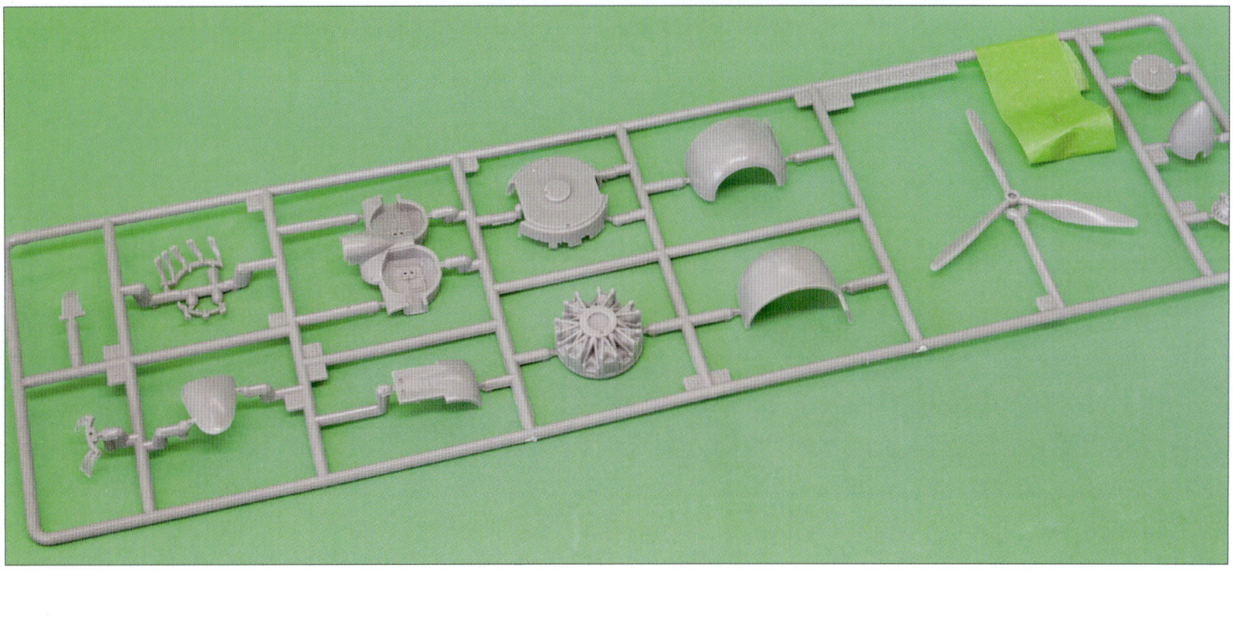

Hasegawa A6M3. (Images courtesy of Frederick Seitz)

Hasegawa A6M3. (Images courtesy of Frederick Seitz)

The latest Tamiya offering is undoubtedly an excellent kit. Key features for me are a very well detailed cockpit plus extensive engine detail with separate cowling plates – the first time this was represented in any scale. In fact, the overall level of detail is probably better than that found in the twice-the-size Trumpeter offering. Although a fairly complex model, assembly is easy and trouble-free. Canopy painting is helped by the inclusion of a pre-printed mask (not pre-cut though) and in all it's hard to see how things could be any better. To be hyper-critical, the cockpit would benefit from the addition of the prominent wiring and plumbing on the floor and sidewalls, and plug leads need adding to the engine. Also, should you choose to leave the cowling panels off, some extra detailing is needed behind them, including representations of the rear engine components and engine bearers. I believe after-market products are available to assist with this if you don't fancy doing the work yourself.

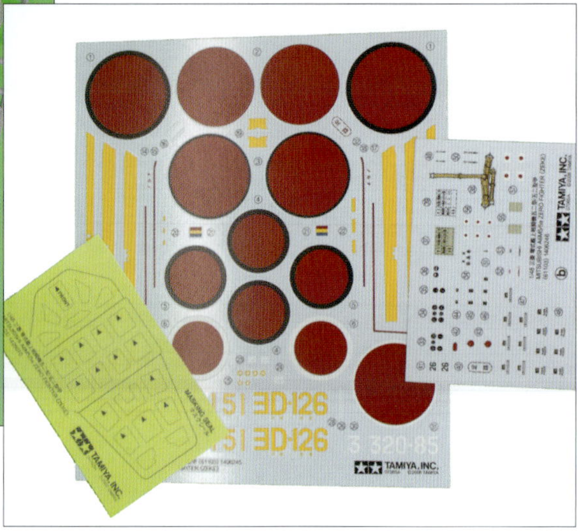

The outstanding Tamiya A6M5.

The completed Tamiya A6M5.
(Images courtesy of Tamiya Inc.)

The Tamiya M3. Tamiya cover all of the main variants. (Images courtesy of Tamiya Inc.)

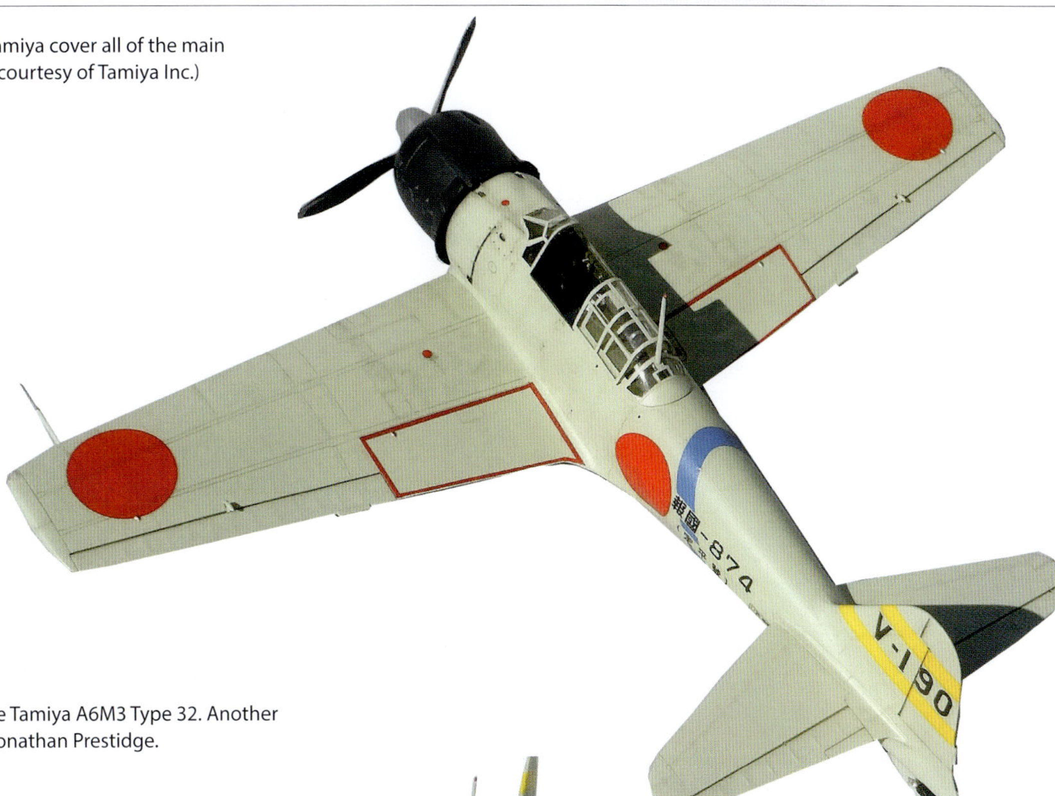

An example of the Tamiya A6M3 Type 32. Another super model by Jonathan Prestidge.

Tamiya A6M3 Type 32.
(Images courtesy of Jonathan Prestidge)

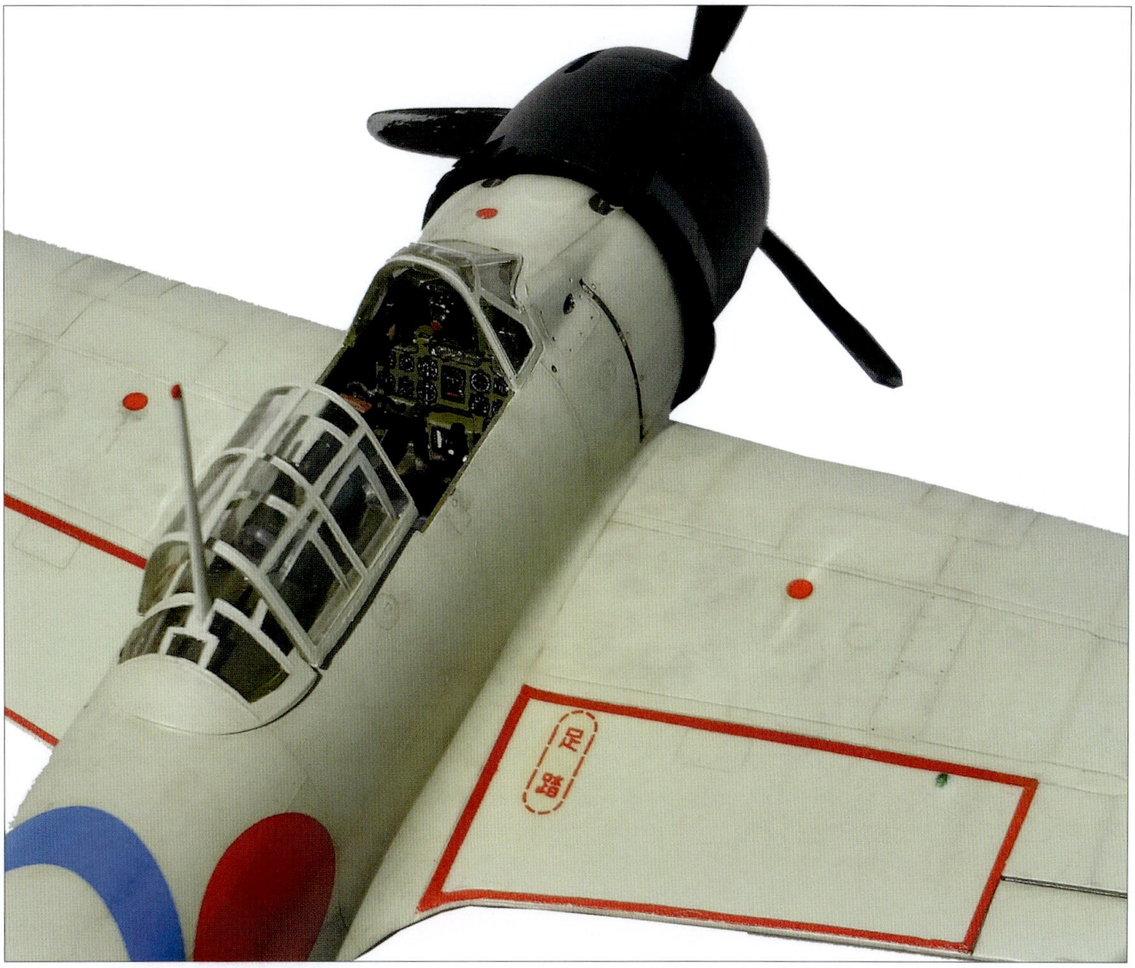

Tamiya A6M3 Type 32.
(Images courtesy of Jonathan Prestidge)

Tamiya A6M3 Type 32.
(Images courtesy of Jonathan Prestidge)

Tamiya A6M3 Type 32. (Images courtesy of Jonathan Prestidge)

Tamiya A6M3 Type 32.
(Images courtesy of Jonathan Prestidge)

When comparing the latest products from these two giant manufacturers, it's important to try and keep a balanced view. The Hasegawa Zeros are very good kits and I would happily build any of them. The Tamiya is more detailed and complex – not everyone wants this, of course – and is more expensive. I also feel it needs some extra work to get the absolute best out of it. The choice is yours.

Moving up to 1/32 scale we first meet the **Revell** A6M5 of 1969. I have fond memories of this kit, having built one as a youngster when it was first released. Even after all this time I can still remember how much I enjoyed it. It was reissued in various re-boxings over the years and although long out of production, is still seen around on the second-hand circuit fairly frequently. In terms of accuracy, span and length are a touch short. Wing and tailplane shape is good in plan view but the fuselage appears a shade pinched aft of the cockpit. The shape of the rudder is not correctly scribed, the droptank is undersized and the spinner too blunt (which largely explains the discrepancy in length). The sliding canopy has overscale guide rails and a rear section that is too shallow, although the canopy components are quite cleanly moulded. The instrument panel is good but the rest of the cockpit detail is either inaccurate or non-existent. Undercarriage bays contain some detailing but the wheels and undercarriage legs are very basic. The main problem comes at the front. The cowling is quite well shaped, but the gun ports are not correctly represented. The cowling gills are a separate moulding and are a very poor fit. The exhaust stubs are too big, incorrectly shaped, and protrude too far from the fuselage. The engine is quite well detailed by the standards of the time but is undersized and has an incorrect crank-case shape. Overall, not a bad effort for its time, if a long way short of definitive by the standards of today.

Coming a year or two later, the **Hasegawa** A6M5(c) is a much more impressive effort. It's actually an A6M5 and would need some modification to make it a (c), but it was still a very good kit for its time. In terms of accuracy, there's not much to fault, but the ailerons on the upper surface of the wings are not correctly scribed. Tailplane shape is generally good, but the tips are a shade too rounded. The fuselage profile is good but the upper section in front of the cockpit is sloped a bit too steeply – strange how this seems to be a bogey area for so many otherwise good kits – but the tricky cowling shape is good, as is the three-part canopy which thankfully is not designed to operate. The engine is a good effort too, with individually moulded cylinders and exhaust pipes. All the other major components are well done, but the cockpit and wheel well detail, although accurate as far as it goes, is pretty spartan by modern standards. To build the kit as a true 5(c) the starboard gun port on the cowling needs to be slightly enlarged and the port-side 7.7mm gun breech removed from the top of the instrument panel. The starboard-side breech needs replacing with a representation of the larger 13.2mm weapon that replaced it. Representations of armoured glass are required behind the windscreen and on the pilot's headrest. Also, the fire extinguishers need adding to the cockpit floor and the seat needs reshaping to the late pattern of the M5(c). The kit has been through a few re-boxings and reissues in its time, including by the Frog company in the UK. It is still frequently seen around and for those prepared to put in the work, provides a budget alternative to the more modern and much more expensive Hasegawa and Tamiya offerings.

Two other 1/32 kits are worth a mention before we get to the 'jewels in the crown'. In around 1988 the Japanese company **Swallow** released an A6M2 Model 21. This was never that widely available in the UK and is a real rarity now, but it was essentially a good kit, pretty much on a par with the previously described Hasegawa offering and having the advantage of offering a different variant. In terms of accuracy, it's really rather good, and despite a fairly basic level of detail, it's worth considering if you happen to come across one at a reasonable price. The other is something of a curio. Appearing in 2007 as one of a small range of 1/32 kits, the A6M3 Model 22 was made in China under the banner of **21st Century Toys**. I bought one because it was very cheap by 1/32 kit standards, and also was the only representation on the M3 Zero variant in that scale. The kit is unusual in that the parts are all packaged separately and there are no sprues. In addition, the main components are screwed, not glued together, with covering caps provided for the fuselage sides and wing surfaces to conceal the screw heads. The general accuracy of the kit is good, compromised mainly by the operating undercarriage and canopy, although the latter is not too bad. The gun ports on top of the cowling are too big but otherwise things check out okay. Cockpit detail is quite good and there is a remarkably detailed and well-moulded pilot figure included. A good quality decal sheet is another positive, with markings for three separate aircraft. The downside is the poor quality of the overall moulding and the surface detailing. The mould quality is way short of the crispness we are accustomed to today. It's more like an old kit from the

The Revel A6M5(c) was one of their earlier 1/32 kits.

Above: Even old kits look good if they're built well. Despite the various faults of the kit, it still makes an attractive model. Another one by Gary Wenco.

1960s where the moulds have got tired. The surface detail is engraved, which is good, but is overdone and lacks sharpness, particularly on the wing surfaces. Nevertheless, considering that this kit can still be sourced, often for bargain prices, it's not one to be dismissed.

Now to the Big Two. The **Tamiya** A6M5 was issued around 2007. In the opinion of many Zero aficionados, this is the definitive A6M kit, a class above anything that has gone before it. Looking through the kit and looking at examples of the finished model, it's hard to disagree. The level of detail is excellent but to me it's the little extras – photoetch seatbelt, canopy painting mask, nicely engineered working undercarriage that is actually done to scale, even the inclusion of a length of fine wire for the aerial cable – that set this kit apart. Even so, the cockpit needs to have the floor plumbing and sidewall wiring added (aftermarket products are available but it's easy to do it yourself with a bit of research), and the otherwise well-detailed Sakae 21 engine needs representations of the plug leads added. Control surfaces are designed to operate and careful engineering means that this feature does not impair scale accuracy, but actually improves it as the hinging mechanism is to scale. Tamiya also produce an A6M2 now. I haven't seen a sample close up but I would expect it to be every bit as good as this one.

Bearing in mind the quality of the Tamiya item, it was something of a surprise that in 2016 **Hasegawa** issued a newly tooled A6M5(c) to replace their worthy effort of some 40 years earlier. Surely we didn't need another 1/32 Zero? In fact, the Hasegawa kit does provide a viable option for, or companion to, the Tamiya effort. In the first place, Hasegawa are offering a slightly different variant. The M5(c) had differences in armament and underwing ordnance provision from the M5. Secondly, while the Hasegawa doesn't have the bells and whistles of the Tamiya, it is still an excellent kit that is simpler to build, while still offering a good level of detail. The

BELOW: The rather quirky 21st Century Toys A6M3 …

… still makes a good model if built sufficiently well. Another of Gary Wenco's.

The completed kit. Photo by permission of Tamiya Inc. This is the M5 – they also produce an M2 in the same scale.

cockpit is very good, though still lacking wiring and plumbing. Engine detailing is less extensive than the Tamiya but is still good enough for most modellers, even if you again need to add the plug leads to the ignition harness. Undercarriage and wheel bays are well done and the standard of finely engraved surface detail could hardly be bettered. The third factor and perhaps the biggest plus for the Hasegawa kit is its price. It is around 30 per cent cheaper than the Tamiya, and sometimes even better than that with some careful online searching. When choosing between these two products, it really is a question of pay your money and make your choice. I would equally happily build either.

Finally, a look at the big boys in 1/24 scale. Actually, until recently it was 'boy' singular, as the **Bandai** A6M5(c) offering was the only show in town. We now have

The amazing – if expensive – Tamiya A6M5.

Some of the range of items in the Tamiya kit that make it a bit special.

The excellent Hasegawa A6M5(c) kit of 2016 replaced their 1970s effort.

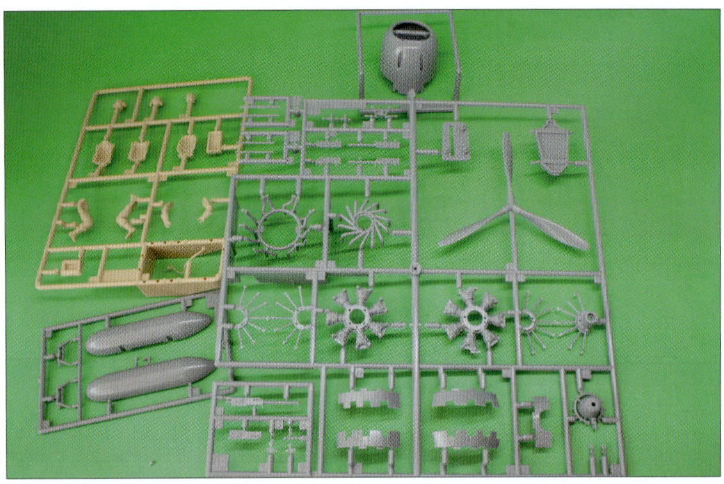

A small sample of some of the parts from the Hasegawa kit gives an idea of the quality of the moulding and detail.

The Bandai 1/24 M5(c) has been around since 1970.

the Trumpeter A6M2 and, I believe, a 'Rufe' as well. The Trumpeter kit is featured in the model showcase, so we'll look at the Bandai kit here, which dates from 1970. It is very hard to find now – and likely to be very expensive if you do. It was a very good effort for its time but doesn't do so well by modern standards. Good points first. The general outline is pretty good. The engine is quite well detailed and is the only kit I've seen to include a representation – inaccurately, it turns out – of the plug leads. A neatly moulded pair of vinyl tyres are included for the mainwheels, and the wing armament has reasonable detail and can be shown off via removable panels. On the down side, the cockpit detail is very basic for this scale. The canopy is quite well shaped but the frame lines are too broad and the operating sliding canopy means the grooves are oversize, although that's less noticeable in this scale than in some of the smaller ones. Surface detail is overdone. It's a mixture of raised and engraved lines. The fuselage isn't too bad in this respect, but the wings are awful, with very deep and not always accurate panel lines and some clunky raised detail as well. The main undercarriage units are designed to be operable and are fairly crude. The most noticeable accuracy issue to me is the front of the cowling, which has a poorly represented intake for the supercharger. It is too big and incorrectly shaped. The position of the gun ports on top are also suspect. It comes with three decal options, although the decals in my sample were cracked and unusable, which is a pretty common problem with old kits. Despite its issues, this kit makes an imposing model even built from the box, and with a bit of work would provide a worthy companion for the Trumpeter M2 or even the Airfix Hellcat.

Acknowledgements

I'd like to thank Bill Anderson, Peter Terry, and particularly Gary Wenco of the IPMS Japanese Aircraft Special Interest Group for their assistance with the production of this chapter.

Thanks also to Pete Binger of The Hobby Company Ltd. for arranging the use of the photos from Tamiya Inc. Finally, my good friend Terry Foley for sourcing the Hasegawa A6M8 kit and the excellent reference book from Japan.

Good and bad points of the kits are evidenced here. A good level of engine detail but overdone surface detailing. Note also the over-scale slide rails for the canopy.

Gary Wenco demonstrates that good building can make the best of a limited kit. This model has real presence.

The Trumpeter 1/24 A6M2, featured in the model showcase section, is a sound if not spectacular kit.

The kit has been available for several years. This one was bought direct from China. The spread of parts is impressive. There are just under 200 altogether.

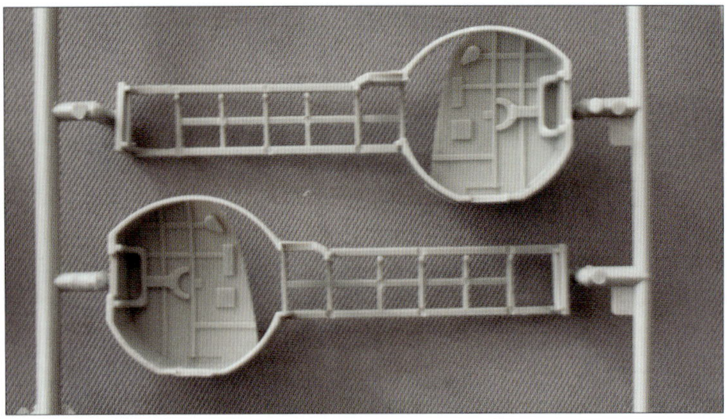

A close-up of the wheel bay shows a fairly basic level of detail.

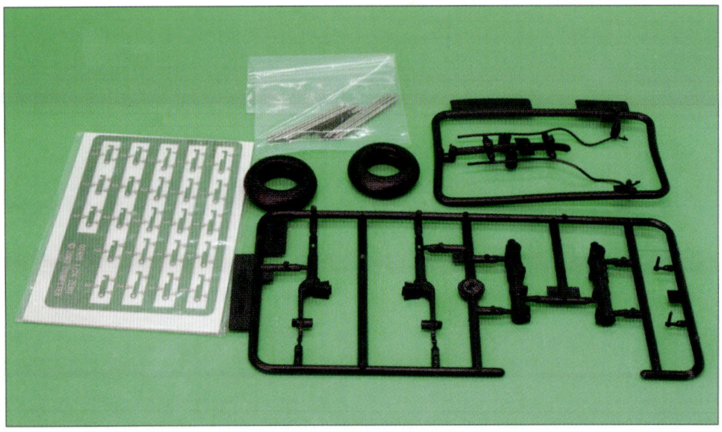

Vinyl, metal and HBS components also feature.

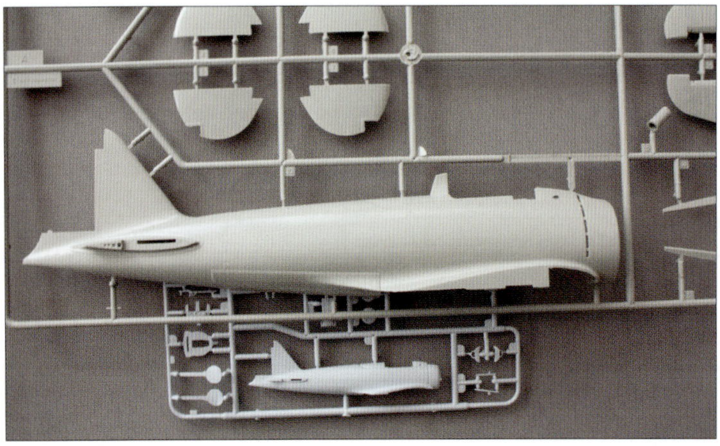

Shown next to the Airfix 1/72 kit parts the large scale becomes apparent.

Model Showcase

Although this kit has been around for a year or two now, I have yet to see a completed example. As I have never built an aircraft kit in this large a scale before and it appears less well exposed than some of the recent Japanese 1/48 and 1/32 kits, it seemed an interesting choice as a showcase build.

Once the large box is opened, the size of the major components is immediately apparent. This is going to be a big model. There are 193 plastic parts, plus a fret of etched metal hinges and lengths of steel rod for the operating control surfaces. Tyres and brake pipes are moulded in black vinyl and the undercarriage legs in a black plastic that I think is ABS – it certainly isn't normal polystyrene. Mouldings are flash-free and surface detailing is generally good, with the distinctive riveted finish of the Zero being captured rather well. On the downside, the general level of detail is no better than on the latest 1/32 scale kits and the sharpness of moulding is not quite as crisp as I found in these smaller scale products. Most annoying of all are the large number of ejector marks found all over the place. Often these are in areas that can't be seen on the finished model, but too often they are in open view and need attention at some stage during the building sequence.

One thing you need to be aware of before undertaking this project is that very little is available in the 'after-market' department should you wish to enhance it in any way. There is a replacement instrument panel – a pointless addition in my view as the kit item is perfectly good enough – and some alternative decals. Whilst I was completing the build, one or two other things appeared but otherwise, that's it. If you want a 'resin this' or 'photo-etched that' or '3D other' you'll be disappointed, at least at the moment. The extra details I included during the course of this build all come under the category of DIY. At least it keeps the cost of the project down!

The breakdown of the parts in this kit is pretty orthodox, and so is the construction sequence. We begin with the cockpit. Some 40 or so components go to build this and it assembles easily. Checking with the various references, of which there are plenty online, it is accurate and generally complete. I thought the instrument panel looked effective but some of the other components, whilst acceptable, lacked the extra crispness of moulding that I alluded to earlier. Also, most of the wiring and pipework on the floor and side consoles has been missed and adding this was the main extra task involved at this stage of the project. Again, easily obtainable photo references make this task reasonably easy. To be fair to Trumpeter, this task is also necessary in the 1/32 Japanese rivals, which don't include the 'plumbing' either.

I had to address three issues in the cockpit though. The first two related to the seat.

The seat itself is accurately shaped but shows some very prominent ejector marks on its visible surface that need to be filled. The compound curve of the seat doesn't make this particularly easy but it has to be done. Equally annoying is the absence of any sort of seatbelt. This is a strange omission because it is such an obvious feature and a scratch-built replacement is required. The large scale of the model means this needs to be a detailed and carefully built item and it took me ages to get a result that looked right. The third problem is the positioning of the machine guns on top of the instrument panel. I initially left these off to help with painting but when I came to fit them I found that, if fitted into the slots provided, they project too far back into the cockpit. The starboard gun can't actually be fitted at all because it fouls the existing detailing on the sidewall. The guns need repositioning some 5mm further forward – again, reference photos are a great help here.

Next comes the engine assembly. As with the cockpit, a goodly number of parts are provided and the general level of detail is good. The main addition necessary is a corrected ignition harness and a full set of plug leads. This is important because the cowling is easily removed from the completed model and all of this extra work will be seen. This is in contrast to a lot of the engine components, which when assembled and placed in the fuselage look initially impressive, but are completely covered later when the fuselage halves are joined ...! A study of the photographs will show what I mean. To display the engine detail behind the cylinder rows you would need to cut away the forward fuselage along the lines of the access panels, and then add details such as fasteners and wiring. I considered doing this but found suitably detailed references hard to come by, so saved myself some work and completed this stage as per the instructions.

To get back to the plug leads, I first modified the inaccurate ignition harness provided in the kit. The 14 small spokes need cutting off and replacing with thicker items, pre-drilled to accept the plug leads. This is a fiddly job and needs a lot of care but the results justified the effort. Next I worked on the cylinders. The two rows were each assembled and cleaned up. Each cylinder had two spark plugs, one at the top front and one exactly opposite at the top rear. These are marked as small blobs on the kit parts, not very convincing but still useful. They provided a drilling point for a 1mm drill bit going through the cylinder at

The basic components of the instrument panel go together quite well.

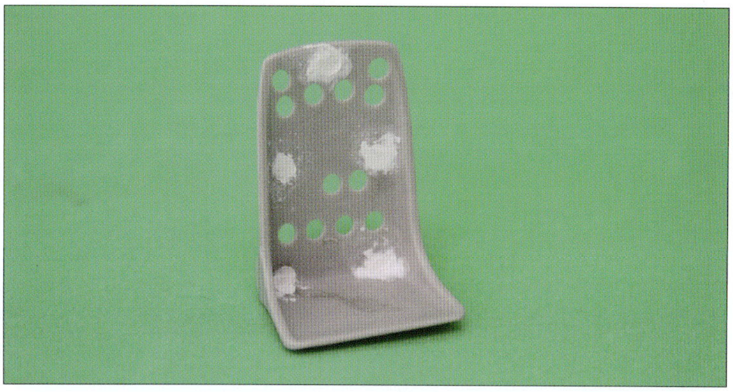

The seat is well shaped but comes with some of the annoying ejection marks that are an unfortunate feature of many of the kit parts.

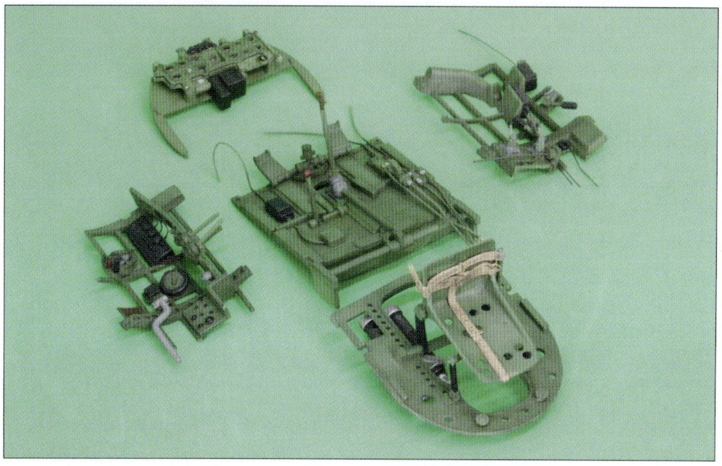

The completed cockpit units ready for assembly. The plumbing and cabling needs adding and a scratch-built seatbelt is also necessary.

Far right: The assembled cylinder banks being drilled for the plugs and leads.

Right: The completed front end of the engine showing the modified ignition harness and the plug leads.

Engine and cockpit assemblies in place in the fuselage. Everything fitted well.

Unfortunately, a large part of the engine disappears from view once the fuselage halves are assembled.

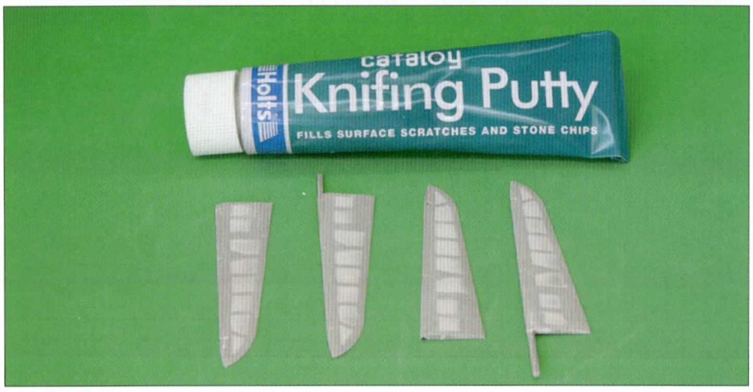

A problem comes in the form of the awfully overdone recessed detailing on the fabric-covered ailerons and elevators ... and the solution comes in the form in the form of Holt's Knifing Putty.

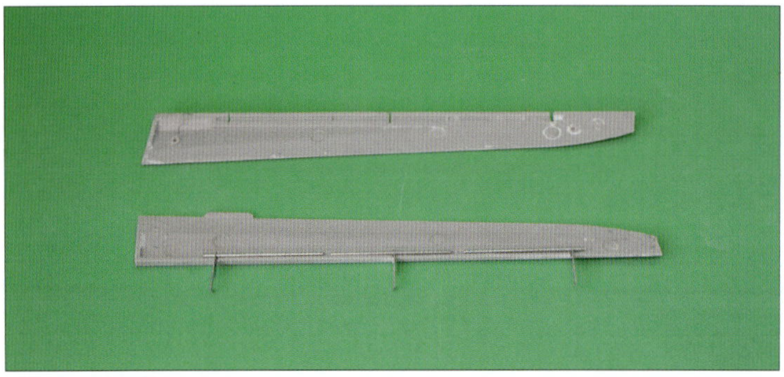

This shows the metal rod and hinge mechanism of the 'working' control surfaces.

the point of the front plug at coming out through the point of the rear one. I then inserted 9.5mm lengths of brass tube through each hole. This stands proud of the cylinder face to provide a representation of the plug. Through the tube can be passed lengths of round section rubber which in turn are glued into the ignition harness. The rubber has just the right amount of flexibility to represent the curve of the plug leads. The leads were added after all the painting was completed, incidentally.

The engine is generally a good feature of this kit. It assembled well, with a good fit of parts and a pleasing level of detail. However, I found a lot of care was needed to get everything lined up accurately. The cylinders, push rods, cowling rings and the intake and exhaust manifolds all have to be positioned correctly and a lot of careful dry-fitting was necessary to ensure this happened. Even after all this, I found the forward crank case cover (that's the dome-shaped thing in the middle of the front cylinder block) did not line up correctly and I think this is actually a fault of the kit because I'm sure everything else was assembled properly. However, it was an easy job to correct it and no harm was done. The engine bearers are the only parts that don't fit well. They need a certain amount of cutting and trimming to get them to length and but for the fact they are out of sight when the build is complete I would have replaced them with suitable lengths of brass rod. The cowling is assembled at this stage. It is well moulded and split along the appropriate panel lines should you wish to display it open. The separately moulded cowling gills are nice and thin, but carry ejector marks on their inner surfaces and are in the open position, with no option for them to be closed. Some careful surgery will be required if you want to model them closed – some optional pieces would have been a useful addition here, I think.

Fuselage assembly comes next and poses no real problems. The engine and cockpit assemblies fit in well. The rudder and tail-wheel are assembled and inserted before the fuselage halves are joined. The tailwheel in particular is a neat little sub-assembly that retracts in the correct manner if you want it do so. A bit pointless though, seeing as the main undercarriage can only be assembled in the 'down' mode. The rudder is also designed to operate – more of this later – but I chose to have it fixed and I suspect most other modellers would do the same. A lot of taping is required to hold the fuselage halves together once they've been glued, but this is purely due to the size of the item and not because the fit is poor. In fact, hardly any filler was necessary once the glue had dried and the seam lines were sanded down.

Wing and tailplane assembly comes next and this is not as straightforward as it first appears. This is partly due to the fact that the aileron surfaces have a grossly overdone rendition of the fabric covering – same applies to the elevators – and this has got to be addressed. I used Holt's Knifing Putty which is nice and fine, bonds well to the plastic and can be sanded to a very smooth finish. Another complication is that the elevators, ailerons, flaps and the folding wingtips are all supposed to be operable. To facilitate this, etched stainless steel hinge plates and pre-cut lengths of steel rod are included. Even though I had no intention of having the items operable on my model, I still used the components and assembled them as per the instructions. They assembled easily enough although cyanoacrylic glue is required for the metal bits. Even though I had all of the operable parts fixed in position, because they follow the same method of construction as the real aircraft, they look very effective when completed and weren't just the gimmick that I first thought.

Wheel bays are represented as separate items and the level of detail is pretty basic for a kit in this scale. I left them as they were, but would seriously consider adding extra detail if I did the build over again. Nevertheless, they fit well and add some welcome strength to the wing assembly. Two Type 99 20mm cannon are included, together with drum magazines and ejector chutes. However, these are buried within the wing and all but the ends of the gun muzzles are completely hidden. The access panels on the upper wings are marked and could be cut out to display the guns, but quite a lot of extra detailing would be required to do this properly and I was unable to source adequate reference material to do this. Once the control surfaces, wingtips and wheel bays are fixed in place on the single piece lower wing, the upper halves can be added. As with the fuselage, there were no significant fit problems although a lot of taping and clamping was required to hold everything in position.

It was now time to add the wings and tailplane to the fuselage. Whenever I build an aircraft model I always find this one of those 'hold your breath' stages, particularly with such a large model as this. In fact, things went quite well. A dry fit looked promising and once the wing was glued in place and given time to dry there was only a small amount of filling necessary, most significantly on the area where the underside of the wing root blends with the fuselage. Tailplanes fitted reasonably well too, but I had to be careful to get them lined up correctly. They have no dihedral and needed a bit of trimming at the root to get this evenly represented on both sides.

Next comes the undercarriage. The mainwheel legs are designed to have a 'working' suspension with a spring/pin arrangement to allow this to happen. Seeing no point in this, I assembled them in a fixed position. This meant I had to take care to get the right degree of compression for a ground-sitting aircraft and a look at

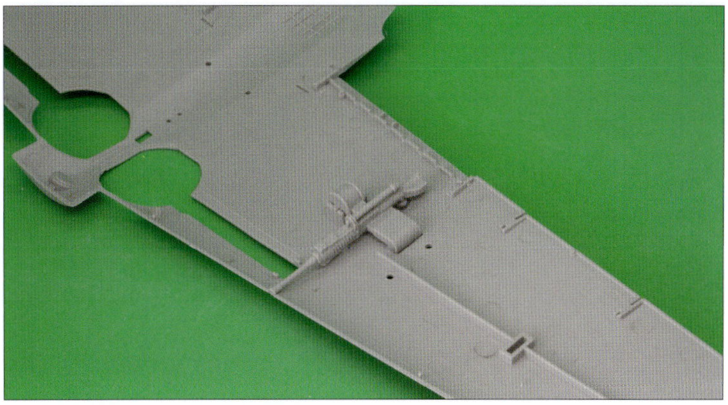

The wing cannons are another example of some rather nice detail that disappears from view on the finished model.

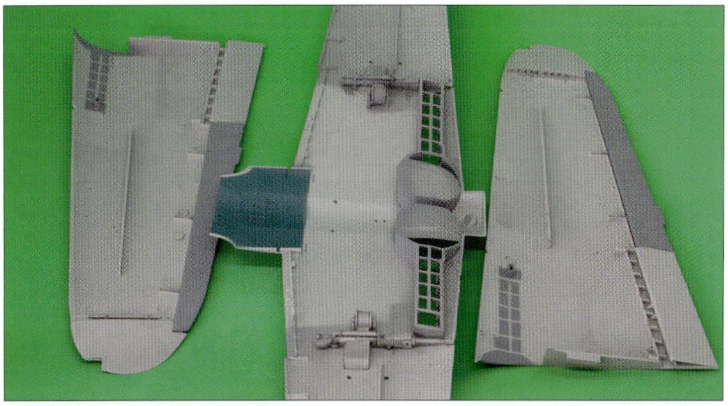

The basic wing units ready for assembly.

Clamps and tape hold everything in place. The fit of the parts was good – it was only the size of the components that necessitated this much securing.

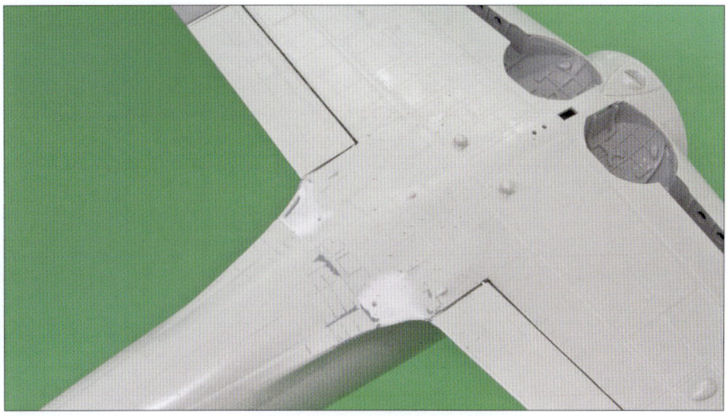

Wing/fuselage fit was pretty good, with just this area here needing a bit of filler.

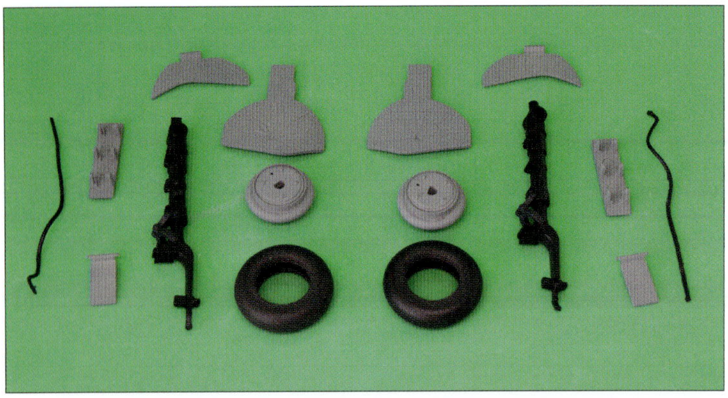

The partially assembled undercarriage units. I liked the vinyl tyres and brake pipes.

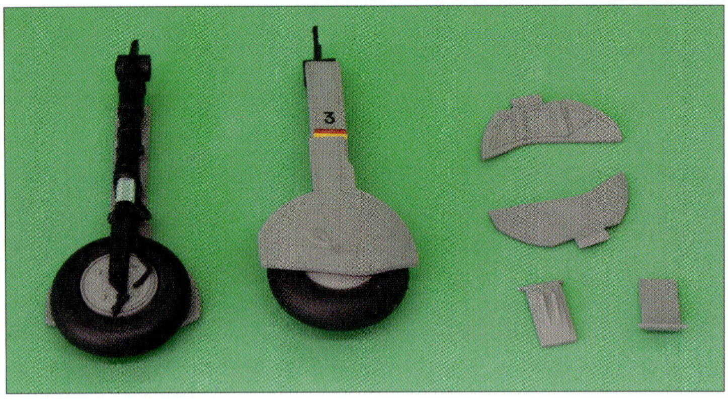

The main undercarriage assembled. It was at this point I first used the kit decals and was not that impressed, although they didn't work too badly here.

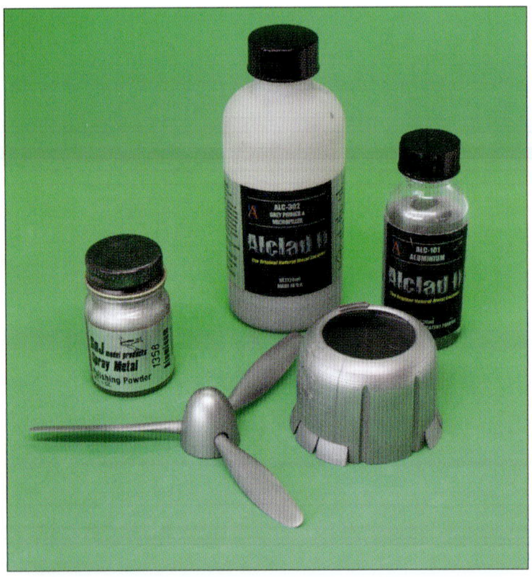

The assembled propeller and cowling after their metallic paint had been applied.

The cowling in its final form. It was an easy job to chip away the softer, dark acrylic to reveal the stronger metal coat underneath. I only did this lightly but could have done much more if I'd wanted.

reference photos was necessary. I also had to ensure that both completed legs were the same length! Vinyl tyres are provided in the kit, and these are quite the best example their type I've come across. The mould seam, so difficult to remove from vinyl items, is almost invisible and they fit well over the wheel hub, saving a bit of fiddly masking. The tyres were buffed with fine wire wool before being put on the hubs as this took the glossy finish off them. Each undercarriage leg is provided with a vinyl brake pipe. These fit well and look effective, although one did have some flash along its length which was not easy to remove. The mainwheel doors are well shaped and fit snugly to the undercarriage legs. They are marred by having ejector marks on their inner surfaces though, which needed attention. The completed units fit securely into the wheel bays and give solid support to the finished model.

I found assembling the propeller to be a trickier job than expected. The first issue is that the attachment points on the sprue for the three separate blades are very thick and leave a fiddly cleaning-up job which must be done carefully as the blades have a metal finish. As we all know, this will ruthlessly show up any surface imperfections so I had to make sure there weren't any. The next issue was that, when assembling the blades to the hub unit parts, the alignment was not particularly positive and I had to make a simple but carefully measured jig to ensure this was right. Once everything was completed and the spinner added, I gave them a metallic finish, which was also applied to the cowling. I used Alclad black primer and aluminium paint for this, finished of with a buff of SnJ aluminium powder. Alclad keys very well into the plastic surface and had a hard, strong finish. This is important with the cowling as it meant I could coat its outer surface with a black/metallic blue acrylic mix, then scrape through the soft acrylic with the point of a cocktail stick to show the metal underneath when I wanted a bit of subtle weathering. The small red stripes and i/d number under the cowling are provided as decals. I found them thin but brittle, and had to touch up the propeller stripes with red paint and a fine brush.

Smaller parts get dealt with now. The drop tank assembles easily and was prepainted before being attached to the model later. Same applied to the bombs and rather clunky bomb racks. These look a bit basic but again I was unable to find adequate references to improve them. The simplest solution is to leave them off altogether, as was more often than not the case with the real aircraft. You will need to fill in the attachment holes though. I decided to include them for the sake of completeness. The oil cooler is a three-part assembly that fits under the fuselage behind the cowling. The intake vane needed a bit of filler but the completed unit was a good fit to the fuselage.

The canopy is the next task. The three pieces are well moulded and commendably thin. A dry fit showed that they mated quite well with the cockpit surround. The frame lines are clearly marked and have a frosted finish, which improves their opacity when painted. When it comes to masking, this is a time-consuming DIY job as no commercial pre-cut mask set was available at the time – this has now changed (see below). The large scale of the pieces makes this job less of a fiddle that it could be though. A problem with the canopy became evident once I had completed the painting. Some of the frame lines look too broad to me. This particularly applies to the horizontal ones. As the frame lines are raised and frosted, this is not an easy problem to solve. If I'd had the time, I would have either plug-moulded a new canopy, using the kit items as a master, or sanded and polished off the etched framework and scribed my own more accurately. This would be a job worth doing as it would significantly improve the appearance of the finished model. I don't expect a nice convenient after-market replacement will be appearing anytime soon!

The final paint job is relatively simple because it's just a single colour. I used Halford's grey plastic primer first, then after a bit of pre-shading, used Vallejo acrylic (71.312 IJN Medium Grey) for the overall colour. I then used a Mig pin wash (A.MIG-1607) to bring out the panel lines and add a bit of depth to the overall finish. When it came to the markings, I decided to spray most of these on due to their basic simplicity and the large scale of the model. Vallejo 71.003 is an almost perfect match for the hinomarus and red fuselage band. Masking was done with a combination of masking film, tape and masking solution and was definitely worth the extra trouble. I used another Mig pin wash (A.MIG-1605) on the red areas and was pleased with the results, adding interest but not looking overdone. I used the kit decals for the smaller items like the wing stripes and fuel filler caps and these were not good. They were very glossy, quite brittle, and did not adhere well. This was in spite of careful surface preparation and the use of a range solvents and setting solutions. They should have been sprayed on as well – it would have been quicker in the long run as well as giving better results. On the tailfin I masked and sprayed the horizontal yellow stripes (Vallejo 71.002) but had to use the kit decals for the numerals. Again, endless problems resulted. There are several sets of after-market decals available for this kit now and I would certainly consider using them if I did the build over again. As well as hopefully better quality adhesion, they also offer some more attractive colour schemes. Another option second time round would be the full stencil set for this kit – I discovered this just a week too late! It's produced

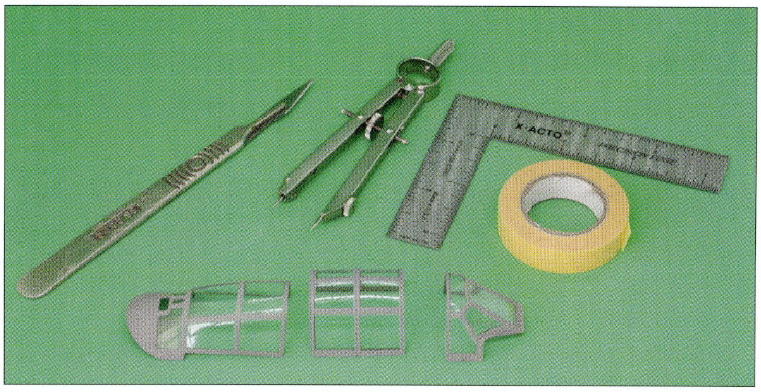

At the time of the build, I didn't know about the canopy masking kit that's now available so I did it the old-fashioned way. The large size of the components meant that it wasn't too much of a chore though.

The model receives its overall coating of Halford's Plastic Primer.

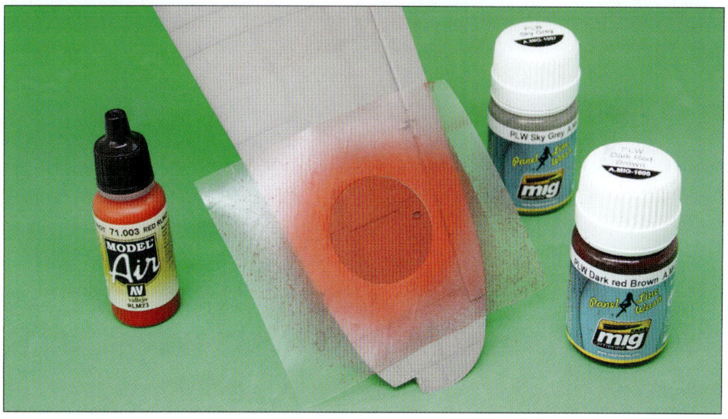

Most of the markings are large and simple enough to be masked and sprayed. The Mig products are used to enhance the surface detail.

Decaling and painting complete.

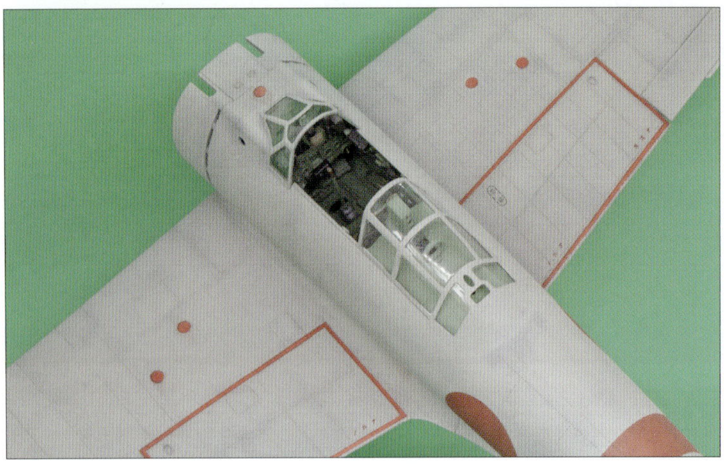

I used the kit decals for the wing stripes and fuel caps. It would have been better to have masked and sprayed them as I had endless trouble trying to get the decals to look right.

by Montex (MT-K24019) and I just wish it had been available at the start!

The final stage of assembly involves fitting the various pre-assembled bits, most of which involve the underside of the aircraft. Bombs and bomb racks, drop tank and arrester hook all fitted well, as did the oil cooler which the instructions have you fit at this stage, although I had actually fitted it much earlier. The undercarriage presented an issue because when the mainwheel legs were plugged into place there was insufficient clearance between the tops of the mainwheel doors (kit parts D30 and 31) and the small outboard door sections (parts D28 and 29). The only solution was to carefully saw off 1.5mm from the top of the main doors. Tricky, because they were in place and couldn't be removed so the job had to be done in situ but it worked.

Whilst fiddling sorting out this problem I encountered another – the ABS plastic used for the mainwheel legs does not respond to normal polystyrene adhesives which means the wheels, doors and brake pipes are, in effect, only a push fit. They do fit well but need careful handling afterwards as they can quite easily come adrift. The final bit of extra work was done with the gunsight. The representation of the Type 98 Reflector Gunsight is correct in term of basic size and outline but crude in terms of detail. I cut away and replaced the transparent sections with squares of clear plastic card and replaced the crude plastic disc that represents the foresight with a photo-etch item from the spares box. As this is quite a visible item the extra work is definitely worthwhile. The radio mast was fitted last, with the wire represented by smoke coloured invisible mending thread.

The model was completed with a couple of coats of Kleer to seal everything and that was that. I had decided at the outset that this model was not going to be heavily weathered, as it represents a carrier-based plane at the very start of the conflict – contemporary photos show how clean these aircraft were at that time. The heavy weathering, such a feature of so many Zeros, comes later in the war. If you do like weathering however, it would be appropriate for this kit if you represent a later war example. A6M2s were around almost to the end and if you source some appropriate markings there will be plenty suitably 'tired' examples to get to work on if that's what you like to do.

So, to conclude, this kit was a straightforward build. It's perhaps not for the absolute tyro but anyone with a bit of experience should have no problem. This ease of construction comes at the cost of a fairly basic level of detail in some areas though. Whilst the kit is presentable built 'from the box', it's not in the same class as the latest 1/32 offerings. However, it's much bigger, doesn't cost a lot more, and has more potential for development. I think an experienced modeller has a lot more opportunity to show off his/her individual skill with this kit than with some of the others where a lot of the work has been done by the manufacturer already.

A lot of good-quality reference material is easily available, and as more after-market stuff is coming on to the scene, anyone building this kit in future has an improving range of finishing options and useful assists to help them along. Perhaps most important of all, this was an enjoyable project. Despite having to build to a deadline, I never got to the stage of having to force myself to go back to it on an almost daily basis. Perhaps that's the most important point of all!

The cowling can easily be removed if you want to show the engine detail.

An obvious area for improvement: the wheel bays really needed more detailing. The colour is correct though. Early Zero wheel bays were painted the same basic colour as the underside, although interior green or *aotake* was also seen.

Various views of the completed model. Despite its limitations, this is generally a good kit and made for an enjoyable project.